Functional Anatomy of the Masticatory System

To my wife Nora and our family

Functional Anatomy of the Masticatory System

W. E. McDevitt BDS, FDSRCS(Edin), FFDRCSI

Head of Division of Restorative Dentistry and Department of Fixed and Removable Prosthodontics, School of Dental Science, Trinity College, Dublin, Ireland; Consultant, Department of Fixed and Removable Prosthodontics, Dublin Dental Hospital, Ireland

WRIGHT
London   Boston   Singapore   Sydney   Toronto   Wellington

Wright
is an imprint of Butterworth Scientific

 PART OF REED INTERNATIONAL P.L.C.

First published 1989

© **Butterworth & Co. (Publishers) Ltd, 1989**

British Library Cataloguing in Publication Data

McDevitt, W.E.
    Functional anatomy of the masticatory system.
    1. Man. Mouth. Anatomy
    I. Title
    611'.31

    ISBN 0-7236-1523-3

Library of Congress Cataloging-in-Publication Data

McDevitt, W. E.
    Functional anatomy of the masticatory system/W.E. McDevitt.
        p.   cm.
    Bibliography: p.
    Includes index.
    ISBN 0-7236-1523-3
    1. Mandible–Muscles–Anatomy. 2. Temporomandibular joint–
Anatomy. I. Title.
    [DNLM: 1. Masticatory Muscles–anatomy and histology.
2. Masticatory Muscles–physiology. 3. Temporomandibular
Joint–anatomy & histology. 4. Temporomandibular Joint–
physiology. WU 101 M478f]
    QM306.M34   1989
    611'.732–dc20

Photoset by Butterworths Litho Preparation Department
Printed and bound in Great Britain by Courier International Ltd, Tiptree, Essex

Acknowledgements

The basic data covered by this review of the functional anatomy of the masticatory system was provided by investigators who recorded and illustrated their research findings and subsequently had their results confirmed by others. The references in the text credit most of the source material, but detailed references may be found in some textbooks that are referred to further.

The reviewer is indebted to many fine teachers for the motivating influences and penetrating understanding that they imparted, especially to the later Professor Ned Keenan of University College Dublin; Professor Robert Last and Dr Stansfield of the Anatomy Department, Nuffield College of Basic Surgical Sciences, London; Dr Lucile St. Hoyme, Anthropologist, Smithsonian Natural History Museum, Washington DC; Dr Richard Koritzer and Peter Neff, Georgetown University Dental School; Dr Rory Mansfield, Department of Anatomy, Trinity College, Dublin; and to Hugh Barry and Bernard McCartan of the School of Dental Sciences, Trinity College, Dublin.

Many of my working colleagues in the School of Dental Science contributed to the production of this text, but I am especially obliged to Patrick Crotty, Frank Houston, Carmel Lyons and Frank Quinn for their informed comment and constructive criticism.

The illustrations are based on originals by Ms Elizabeth Pierce, Ms Nola McNestry and Mr Denis Daly.

Ms Ann O'Byrne, of the Dental School Library, sorted the reference material and advised on the layout and listing of contents.

The task of typing the manuscript through numerous drafts was carried out, patiently and thoroughly, by Ms Ann Golden and Ms Deirdre Byrne.

I am deeply indebted to the Dean of the School of Dental Science, Trinity College, Diarmuid Shanley, and the recently retired Chief Consultant, Professor Rodney Dockrell, for the facilities they placed at my disposal and their interest in and encouragement for this project.

Preface

Functional, applied anatomy is the Cinderella branch of the subject and nowhere is this more true than of the masticatory system. However, the changing pattern of dental disease, the rapid advances in maxillofacial surgery and the modern requirement for good masticatory function has caused a renaissance in our perception of the importance of the anatomy of this system.

The basic science of this subject is carried out in many centres, world-wide. Consequently, it is not easy to keep track of developments in the field and the significance of new insights is often lost to everyday clinical dentistry. Language difficulties and distance cause gaps in communication. Sometimes significant developments are overlooked and forgotten. Thus, it is hoped that this review of the functional anatomy of the masticatory system will add appreciably to the basic descriptive anatomy that is taught, in the traditional fashion, in university departments.

A comprehensive understanding of the anatomical basis of adult mandibular movement and its relationship to dental occlusions is required for all dental disciplines. There are two chief aspects to the topic:

1. The musculature attached to the mandible; its distribution, attachments and coordinated and varying activities as prime mover, synergist or antagonist.
2. The pivoting and guiding activity of the craniomandibular articulation; the detailed structure and functional activity of the joints.

Each of these aspects occupies roughly one-half of the manuscript. Only those aspects of importance to our understanding of function and functional relationships are covered. Without these facts it is not possible to understand basic mandibular positions and movements, or to treat, in a rational manner, the common dysfunctions which arise in the system.

The osteology of the system is dealt with excellently in the standard texts DuBrul (1980), Williams and Warwick (1980). Hence little time is given to 'the bones', except for short commentary on the mandible.

Human craniomandibular articulation is unique. It consists of mutually interdependent right and left parts, each with two joints (four joints in all) that are involved in functional movement. To help understand the complexity of craniomandibular articulation, the fundamentals of joint structure and function are dealt with in a separate introductory section. Only when the first section is understood will one have a proper basis from which to tackle the structural and functional relationships of the craniomandibular articulation proper and gain an understanding deep enough to practice general dentistry.

W. E. McDevitt
School of Dental Science,
Trinity College, Dublin

Contents

Part 1

The mandible and the muscles that move it

Chapter 1

Introduction

A good knowledge of mandibular positions and movements is a fundamental requirement for the practice of modern dentistry. Hence, from the point of view of the functional anatomy of the masticatory system, there is a need to understand:

1. The straightforward mechanical possibilities for the direction of mandibular movement created by:
   (a) the points of attachment of the major groups of muscle fibres;
   (b) the alignment of the major groups of muscle fasciculi and tendons of attachment.
2. How the internal architecture of the mass of muscle tissue may modify:
   (a) the force developed by the tissue;
   (b) the range of movement;
   (c) (perhaps) the direction of movement.
3. That a named mass of muscle fasciculi does not act alone, but with its synergists and antagonists. (This is especially so for muscles moving bones through the three dimensions of space, simultaneously, with weak joint guidance, e.g. those attached to the scapula or to the mandible. Assumptions based on simple mechanical principles of lever action, while probably applicable to the major directions of mandibular movement, may not describe the mechanism involved in the refined, final positioning of the mandible to achieve non-traumatic occlusal contacts.)
4. The continuous nature of the sheet of muscle tissue on the sides and base of the skull concerned with moving the mandible, its developmental source and the many confluences of differently named muscle masses. Since most of this tissue can be dissected off the skull as a continuous confluent mass (Koritzer, 1983), it is probably more useful for understanding function to consider it as a mass of tissue available to the central nervous system for the movement of the jaw as required by the conditions existing at a particular time, rather than as compartmentalized muscle tissue with a specific, but theoretical, mechanical task.

The elevators of the mandible were described by Leonardo da Vinci in the late fifteenth century and by Vesalius in the sixteenth century. The jaw movements were described by Ferrein (1774). The central European anatomists of the nineteenth and twentieth centuries applied scientific method to expand and rationalize the anatomy of the masticatory system, especially the craniomandibular joint (Meyer, 1865; Wallish, 1922). One of the great fundamental investigations into the latter was carried out by Rees (1954) in Belfast, Ireland. In recent times, major advances in form and function have come chiefly from the Scandinavians, particularly in regard to the craniomandibular joint. Finally, electromyography is beginning to reveal the role of the muscle tissue in the various functional and parafunctional activities of the masticatory system.

Chapter 2

The mandible

The human masticatory system is primarily designed to ingest food and prepare it for swallowing.

For the ingestion, crushing, shearing and swallowing of food the upper and lower arches of teeth must be able to separate and move forcefully towards and over each other. The simplest way to do this is to fix one dental arch in position and move the other teeth in relation to it. The human upper dentition is fixed to the skull base. To be made mobile, the lower arch of teeth is fixed to a bone, the mandible, which can be moved and used as a lever to apply force.

Food is crushed and shredded by moving the lower dental arch through it, towards and over the fixed upper arch of teeth.

For swallowing, the mandible must be stabilized at a suitable height in order to elevate the hyoid bone and larynx. This is accomplished by moving the mandible into the maximum intercusping position. The hyoid is then elevated by the suprahyoid muscles.

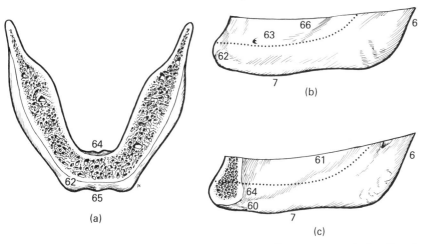

Figure 1 The mandible is basically a U-shaped bone surrounding a neurovascular bundle and supporting various processes: (......) approximate position of neurovascular bundle; (a) superior perspective; (b) lateral perspective; (c) lingual perspective; (6) posterior border; (7) inferior border; (60) digastric fossa; (61) mylohyoid crest; (62) mental tubercle; (63) mental foramen; (64) genial tubercle; (65) symphysis menti; (66) external oblique ridge

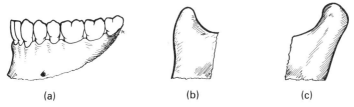

Figure 2 The processes attached to the mandible: (a) alveolar process; (b) coronoid process; (c) condylar process. The mandible supports the alveolar process and divides into the latter two processes

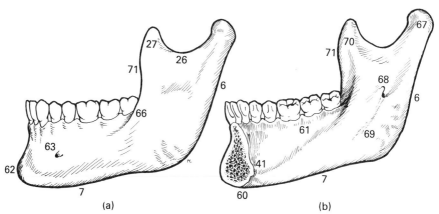

Figure 3 The mandible in lateral (a) and lingual (b) perspective. The basic body of the bone supports the alveolar process and divides into the (i) major muscular process – the coronoid process and the (ii) pivoting process (and fulcrum) – the condylar process. Most of the muscle tissue which moves the bone is attached anterior to the condylar process: (6) posterior border of mandible; (7) inferior border of mandible; (26) masseteric notch; (27) coronoid process; (60) digastric fossa; (61) mylohyoid crest; (62) mental tubercle; (63) mental foramen; (66) external oblique ridge; (67) pterygoid fovea; (68) lingula; (69) groove for nerve to mylohyoid; (70) deep temporal crest; (71) anterior border of ramus

The mandible is basically a U-shaped bone built around a central neurovascular supply (Figure 1). Added to the U-shaped body are various processes (Figure 2) which enable the bone to carry out its primary function of supporting the lower arch of teeth and using it to masticate food. These processes are:

1. The alveolar process, which carries the teeth in a horizontal arch and fixes them to the basic U-shaped bone (Figures 2 and 3).
2. The muscular processes, which provide an attachment area (chiefly for muscle tissue) enabling the mandible to move in the vertical plane and providing the power for mastication (Figures 1–3):
   (a) coronoid process;
   (b) deep temporal crest;
   (c) ramus:
      (i)   anterior border and external oblique line;
      (ii) lateral surface and angle;
      (iii) medial surface and angle;

(d) genial tubercles;
(e) mylohyoid crest.
In addition, the mandible is characterized by attachment fossae (the digastic and pterygoid fossae) for the attachment of muscles to move the mandible other than upwards in the vertical plane.
3. Articular processes which stabilize the moving bone and provide hinging and pivoting surfaces and fulcra for the application of force (Figures 2 and 3):
(a) right and left condylar processes;
(b) mandibular tooth surfaces.

Chapter 3

Prime movers of the mandible

The masseter

‹The masseter is a bulky, rectangular mass of muscle on the side of the face. It is attached between the lateral surface of the ramus of the mandible and the zygomatic arch and is just beneath the skin.

The rectangle is aligned diagonally with one 'angle' very rounded to correspond with the rounded outline of the angle of the mandible (Figure 4).

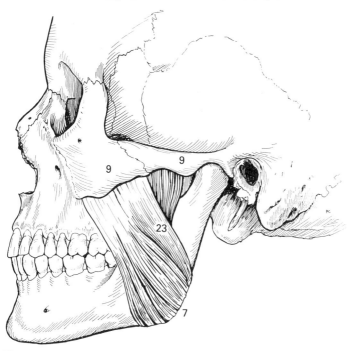

Figure 4 Diagrammatic representation of the superficial part of the masseter as seen in sagittal perspective: (7) inferior border of mandible; (9) zygomatic arch; (23) superficial part of the masseter muscle

   In cross-section the muscle is bulky and wedge-shaped. The thin end of the 'wedge' is the posterior border. The anterior border is much thicker and more rounded, especially in its mid-portion.

   For functional purposes the masseter muscle can be divided into a superficial and a deep part.

The superficial part

This arises from sagittally-orientated, parallel, tendinous septa attached to the lower border of the zygomatic arch, as far forward as the zygomatic process of the maxilla (DuBrul and Menekratis 1981). The most superficial fibres run inferiorly, with a distal and slightly medial inclination (Figures 4 and 5).

   The superficial head fibres are inserted into the lateral surface of the lower half of the ramus and body of the mandible, as far forward as the second molar tooth. Most of the fibres are inserted into sagittally-orientated, tendinous septa with lamellated, irregular, free borders (Figures 7–9; Ebert, 1939). The most superficial

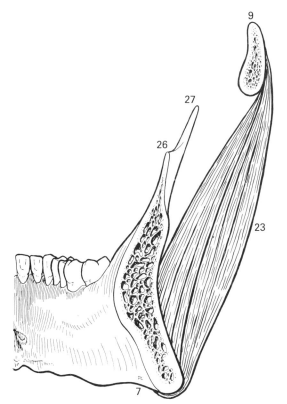

Figure 5 Diagrammatic representation of frontal section through the ramus, zygomatic arch and superficial head of the masseter to indicate the medial-lateral inclination of the muscle fibres and attachment of the masseter to the lateral surface of the ramus: (6) posterior border of cut ramus; (7) inferior border of mandible; (9) zygomatic arch; (23) superficial head of masseter; (26) head of masseteric notch; (27) coronoid process

fibres pass around the lower border and gain insertion on the ramus below the attachment of the medial pterygoid muscle (MacConail, 1975; Figure 5).

In sagittal perspective the origin is wide, extending from the zygomatic bone, anteriorly, to the zygomatico-temporal suture, posteriorly. From this wide origin the muscle converges to a narrower insertion.

When the muscle is well developed, the insertion on the body of the mandible may extend anteriorly to the second molar tooth. This gives the anterior border a concave appearance in sagittal perspective (DuBrul, 1980; Figure 4).

The deep part

This arises from the medial surface and lower border of the zygomatic arch (Figures 6–12). The fibres have a shorter and more vertical course, viewed in the sagittal plane, than the superficial head and are inserted into the lateral surface of the upper half of the ramus and the coronoid process (Kawamura, 1968). The most distal fibres run vertically, downwards, or may have a slight anterior component from origin to insertion when examined in lateral perspective (Figures 8–12). The

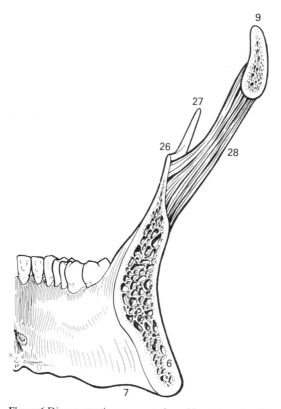

Figure 6 Diagrammatic representation of frontal section through ramus, zygomatic arch and deep head of masseter to indicate the medial-lateral inclination of the muscle fibres: (6) posterior border of ramus (cut surface); (7) inferior border of mandible; (9) zygomatic arch; (26) edge of masseteric notch; (27) coronoid process; (28) deep head of masseter

fibres of the deep head, arising anteriorly from the deep tendinous septum and the zygomatic arch, are aligned, obliquely, in a distal direction, like those of the superficial head, and join up with the tendinous lamellae and lateral surface of the ramus (Schumacher, 1961). Further back the deep head fibre alignment becomes more upright until, eventually, the fibres run vertically, i.e. the masseter fibres present a cruciate arrangement in lateral perspective. The top part of the deep head can be seen in the postero-superior corner of the muscle disappearing vertically beneath the obliquely-aligned superficial head fibres (Figure 8).

When the condyle is protruded the mandibular attachment of the vertical fasciculi of the deep head are carried anteriorly, and these fasciculi are given an anterior inclination. Contraction of these fibres with the mandible protruded will create a retrusive force (DuBrul, 1980).

All of the deep head fibres have a significant medial inclination when seen in frontal perspective (Figures 6 and 7). The muscle fibres arising from the medial of the zygomatic arch and deep layer of the temporal fascia (Figure 12) run a very short course to gain insertion into the edge of the anterior half of the masseteric notch and lateral aspect of the coronoid process (Figures 6 and 7). These fibres

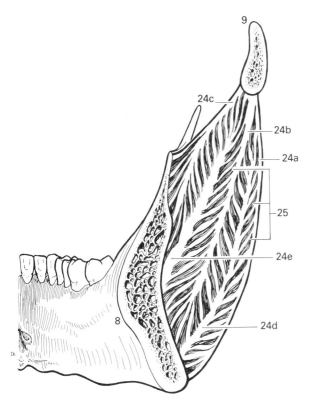

Figure 7 Diagrammatic representation of the tendinous septa of the masseter muscle in frontal section (after Schumacher, 1961): (8) cut ramus of mandible; (9) zygomatic arch; (24a) superficial tendinous septum; (24b) middle tendinous septum; (24c) deep tendinous septum; (24d) lower mandibular septum; (24e) upper mandibular septum; (25) muscle fasciculi

cannot be definitively distinguished from those of the temporal, e.g. by an enclosing muscle fascia. Anteriorly, the masseter and temporal muscles are separated by an extension of the buccal pad of fat.

Tendinous septa

The internal structures of the masseter and medial pyterygoid muscle is dominated by the presence of large tendinous septa (plates, lamellae) which greatly enlarge the surface available for the attachment of the muscle fasciculi (Figures 7–12; Schumacher, 1961).

In the masseter, the commonest finding is the presence of three such septa attached to the zygomatic arch and two septa to the lateral surface of the mandible (Ebert, 1939). The septa below interdigitate with those above (Figures 7–12).

Of the three septa attached to the zygomatic arch, one is superficial and two are deeply placed. All are tough collagenous sheets that are arranged parallel to each other and wide in the sagittal plane (Figures 8–11).

The most superficial of the tendinous septa arises from the anterior half of the inferior border of the zygomatic arch (Figure 8). It extends around the anterior border of the upper part of the muscle to meet the anterior border of the middle septum attached to the zygomatic arch, the two septa forming a fibrous pocket (Ebert, 1939). It extends over one-third to one-half of the upper surface of the

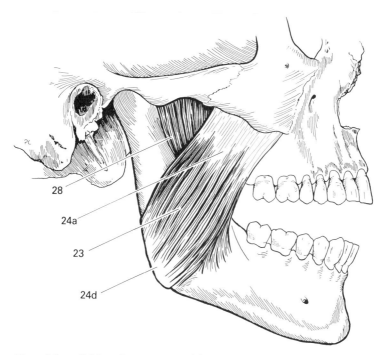

Figure 8 Superficial tendinous septum of the masseter in lateral perspective. The septum is attached to the inferior border of the zygomatic arch in its anterior part. The lower border is irregular and turns medially into the muscle (after Schumacher, 1961): (23) superficial head of masseter; (24a) superficial tendinous septum; (24d) lower mandibular septum; (28) deep head of masseter

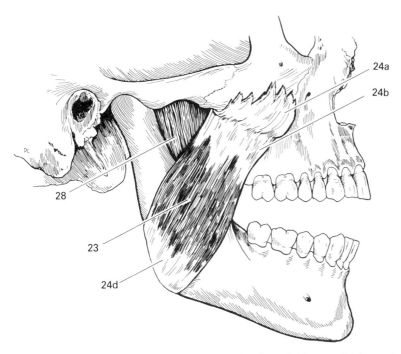

Figure 9 Sagittal view of masseter at a deeper level than figure 5. The superficial layers have been dissected off and the superficial tendon has been reflected upwards. Note that the middle septum extends more distally than the superficial septum. The greatest bulk of muscle tissue is in the middle of the muscle (after Schumacher, 1961): (23) superficial head of masseter; (24a) superficial tendinous septum; (24b) middle tendinous septum; (24d) lower mandibular septum; (28) deep head of masseter

muscle (Figure 8). Its medial surface (Figure 7) is provided with numerous downward extensions, in the direction of the long axis of the muscle, which increase the surface area of the attachment for fasciculi (Figure 8). The lower border has a very irregular outline and is often split into sub-lamellae. Its distal border is usually split and merges into the substance of the muscle to provide attachment from the medial and lateral surfaces (Schumacher, 1961).

The middle septum of the zygomatic arch extends distally, further along the arch than the superficial lamella (Figures 9 and 10). The middle septum reaches the temporo-zygomatic suture and defines:

1. The distal extent of the superficial head of the masseter.
2. The lateral extent of the deep head of the masseter.

The middle septum is less dense and shorter than the superficial tendinous lamella, rarely extending inferiorly below the upper third of the muscle length. Its anterior border may meet with that of the superficial septum, particularly in its upper part, and so form a 'pocket' which is open both below and posteriorly. The lateral side has extensions in the long axis of the muscle providing an extra attachment area for muscle fasciculi. These are not as numerous or as extensive as those seen on the medial side of the superficial tendon.

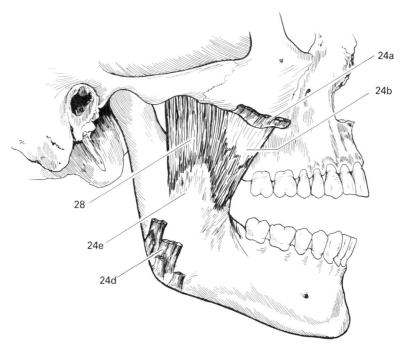

Figure 10 The deep head of the masseter in lateral perspective. Note the gradually more vertical alignment of fibres from front to back and the insertion into the upper tendinous septum of the mandible (after Schumacher, 1961): (24a) superficial tendinous septum; (24b) middle tendinous septum; (24d) lower mandibular septum; (24e) upper mandibular septum; (28) deep head of masseter

The deepest tendinous septum arising from the zygomatic arch marks the origin of the fasciculi of the deep head of the masseter. This lamella extends along the length of the arch to within 1 cm of the craniomandibular joint capsule (Schumacher, 1961; Figure 11).

Usually it is a delicate sheet of collagenous tissue extending inferiorly about 1 cm from the zygomatic arch. The muscle tissue arises from its lateral surface and irregular lower border.

Sometimes its anterior border is continued around the anterior border of the masseter to meet the middle septum.

Two tendinous septa arise from the lateral surface of the mandible (Figure 11). These interdigitate with the upper septa (Figure 7). The mandibular septa are attached to the mandible near the angle and closer to the posterior border, less continuously sheet-like in the sagittal plane and often as an irregular stepwise series of broad tendons (Figures 11 and 12). They are orientated outwards, upwards and anteriorly to lie between the upper lamellae, and usually extend at least half the length of the muscle (Figures 9–12).

It is difficult to define where one of these mandibular septa ends and the other begins. The uppermost defines the posterior border of the deep head of masseter and provides insertion for the deep head fibres. The lower is ideally suited to accept insertion of the fasciculi arising from the medial surface of the superficial lamella.

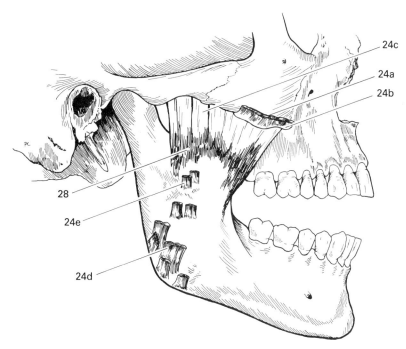

Figure 11 Deep head of masseter in sagittal view, showing the deep tendinous septum. Note attachment of septum to almost the entire length of the zygomatic arch and the short length of the muscle fasciculi: (24a, b, c) superficial, middle and deep tendinous septa; (24d, e) lower and upper mandibular tendinous septa; (28) deep head of masseter

Significance of tendinous septa

The tendinous septa greatly increase the surface area available for muscle attachment, allowing the masseter to have a multipennate structure. This vastly increases its power and may enlarge its options for directional pull. As a result the major bulk of masseter tissue is in the middle of the muscle.

Function of the masseter

The masseter is a bulky, powerful muscle used when heavy crushing and grinding effects are required to chew tough foods.

As well as elevating the mandible vertically, the masseters can exert a range of anterior vectors to the jaw as it is elevated from a depressed position towards the maximum intercusping position on the mid-sagittal arc of elevation. This is accomplished by the spread of contraction from the most anterior to the most posterior fasciculi (Kraus, Jordan and Abrams, 1969). The deep head can exert a retrusive effect (Sicher, 1951).

It is possible that the powerful superficial head could impart a significant anterior component to the mandible close to the centric jaw relation position (Kawamura, 1968; Williamson, 1980).

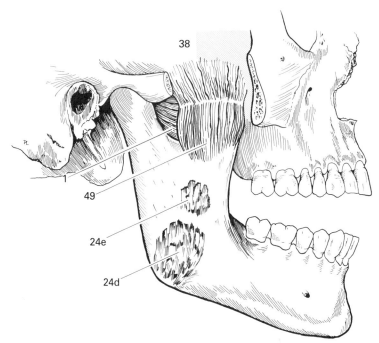

Figure 12 Sagittal view of deepest part of masseter. The fibres arise from the medial aspect of the zygomatic arch and the temporal fascia: (1) lateral pterygoid fascia; (24d) lower mandibular septum; (24e) upper mandibular septum; (49) zygomatico-mandibular muscle; (38) buccal pad of fat extending into temporal fossa

Blood and nerve supply to the masseter
The masseter branch of the fifth cranial nerve supplies innervation. The arterial suply comes from branches of the masseteric artery.

Summary
The masseter is a bulky, rectangular muscle superficially placed between the zygomatic arch and the ramus of the mandible.

Combined action of the superficial and deep parts will seat the ipsilateral condyle into the close-pack position in the craniomandibular joint, with a supero-anterior thrust. The medio-lateral fibre alignment enables a lateral component of force to be developed which may be of significance. The multipennate arrangement of the fasciculi increases the power capability and, perhaps, the variability of lateral movement. The deep part can develop a retrusive component.

The medial pterygoid

The medial pterygoid is a thick, rectangular, powerful mass of muscle tissue on the medial side of the ramus of the mandible. It is not as wide or as thick as the masseter. Its posterior border aligns with that of the masseter in lateral projection, but its anterior border is more dorsally placed (Schumacher, 1961).

In horizontal section, the upper half of the medial pterygoid is wedge-shaped with the thin edge towards the back. The lower half is oval-shaped.

Origin of the medial pterygoid

Most of the muscle arises from the medial surface of the lateral pterygoid plate, the pterygoid fossa and pyramidal process of the palate bone. However, a smaller head arises from the distal surface of the maxilla superior to the level of the buccinator–superior constrictor (Figure 13).

Course and insertion of the medial pterygoid

The fibres from both heads run inferiorly, with definite and significant lateral inclination (Figures 14 and 15a). The latter feature is often omitted by descriptive anatomists who tend to present the muscle in sagittal section. The fibre alignment in coronal section is important to our understanding of the function of this muscle (Figure 15a).

The two heads of origin of the medial pterygoid embrace the inferior part of the origin of the lateral pterygoid muscle (Figures 13 and 14) and mask digital access to the lateral pterygoid in this region.

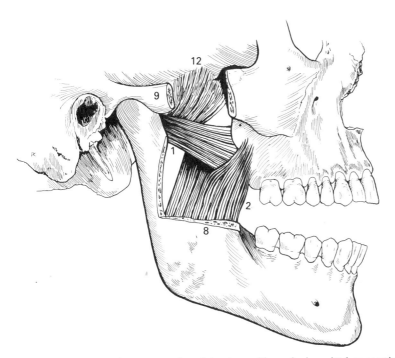

Figure 13 Diagrammatic representation of the pterygoid muscles in sagittal perspective. The deep position of the medial pterygoid can be appreciated together with the antero-posterior alignment of the muscle fibres from origin to insertion: (1) lateral pterygoid, inferior part; (2) medial pterygoid; (8) cut ramus of mandible; (9) zygomatic arch (cut); (12) lateral pterygoid, superior part

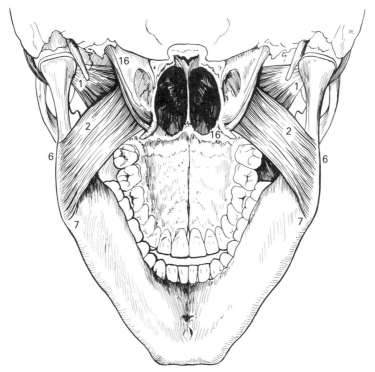

Figure 14 Diagrammatic representation of the pterygoid muscles as seen from postero-inferior view. The medial to lateral alignment of the medial pterygoid muscle fibres is evident: (1) inferior head of lateral pterygoid; (2) medial pterygoid; (6) posterior border of ramus; (7) inferior border of mandible; (16) pterygoid plates

The internal structure of the medial pterygoid

Like the masseter, the medial pterygoid has a complicated pattern of internal tendinous septa which provides a large surface area for the attachment of muscle fasciculi. These are best appreciated in frontal section. Three main septa, attached to the pterygoid structures, provide a base for the muscle fibres, while three septa attached to the postero-inferior medial surface of the mandible provide an insertion area. The pterygoid and mandibular tendinous lamellae interdigitate with each other (Schumacher, 1961; Figure 15).

As in the case of the masseter, the tendinous septa enable a multipennate structure of muscle fasciculi to be developed. This increases the functional ability and power of the medial pterygoid.

The primary function of the medial pterygoid is to elevate the mandible. An anterior component may be given to the mandible by this muscle near the maximum intercusping position (Williamson and Brandt, 1983). Bilateral action of the muscles will seat the condyles antero-superiorly, in the close-pack position in the craniomandibular joint.

Besides elevating the mandible, the unilateral action of the medial pterygoid muscle can impart an anterior and, maybe, a significant medial vector to

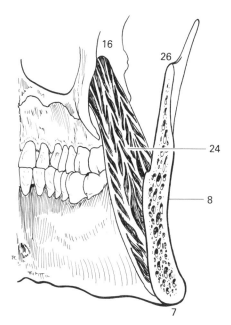

Figure 15 Schematic drawing of tendinous septa of the medial pterygoid muscle in frontal section. Three septa arise from the pterygoid fossa and three from the mandible, and these interdigitate with each other. Sub-lamellae arise from the main septa and the whole system greatly increases the area for the attachment of muscle fibres. Nevertheless, the medial pterygoid is a much less bulky muscle than the masseter (after Schumacher, 1961): (7) inferior border of ramus; (8) cut ramus of mandible; (16) pterygoid fossa; (24a, b, c, d, e, f) main tendinous septa; (26) edge of masseteric notch

mandibular movement. Many believe that the medial pterygoid is the chief source of power for the 'immediate' mandibular lateral translation (McCollum and Stuart, 1937); if such a movement occurs. Sicher and DuBrul dispute the latter point (DuBrul, 1980; Figure 16).

The medial pterygoid is innervated by the medial pterygoid branch of the mandibular division of the fifth cranial nerve. Its blood supply is derived from the medial pterygoid branches of the maxillary artery.

The accessory medial pterygoid

This is a discrete, flat, triangular sheet of muscle tissue on the medial side of the medial pterygoid muscle (Koritzer and Suarez, 1980).

The base of the triangle represents its origin on the skull, extending from the pterygoid plates, anteriorly, along the medial side of the foramina ovale and spinosum, almost to the carotid canal. The fibres converge, inferiorly, to gain insertion into the medial surface of the medial pterygoid muscle 25 mm below the skull. The insertion is dentritic in nature and cannot be blunt dissected (Koritzer and Suarez, 1980).

The accessory muscle is present as an easily identifable entity in over 80% of cases.

The function of the accessory medial pterygoid muscle is not known but it is though to have proprioceptor–myomonitor activity in relation to the action of the medial pterygoid, i.e. it provides detailed feedback to the central nervous system in relation to the activity of the medial pterygoid. This assists refinement of the action of the medial pterygoid, especially as the teeth approach occlusal contact (Koritzer and Suarez, 1980).

The muscle is innervated by a separate branch of the mandibular division of the fifth cranial nerve and is supplied with blood by branches of the maxillary artery.

The combined masseter and medial pterygoid

The masseter and medial pterygoid together, on each side, form a V-shaped sling which suspends the mandible from the sides and base of the skull. The masseter and medial pterygoid attachments to the mandible are closely related. The usual description states that they are confluent at the inferior border of the mandible. MacConail (1975) points out that the superficial head fasciculi of the masseter pass around the inferior border to gain intraosseous insertion on the medial side of the ramus, at and below the line of insertion of the medial pterygoid muscle (Figure 5).

These two muscles act synergistically as the power elevators, driving the teeth through tough food and giving the power for crushing and grinding.

As well as moving the mandible vertically, the medial pterygoid muscles, both individually and collectively, give an anterior component. This tends to seat the ipsilateral condyles in the close-pack position in the joints, but also renders the identification of a clinical reference position difficult. The medial pterygoid, acting

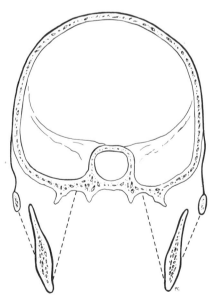

Figure 16 Frontal section through skull and elevators. In this plane the masseter makes an angle of 25 degrees and the medial pterygoid an angle of 15 degrees with the perpendicular

unilaterally or in combination with the contra-lateral masseter, probably can impart a medial thrust to its own side of the mandible, i.e. it can assist the ipsilateral lower head of lateral pterygoid to execute a lateral movement of the mandible (Piersol, 1923; McCollum and Stuart, 1937; Kawamura, 1968; Koritzer and Suarez, 1980; Figure 16). This is disputed by Sicher (1960).

Anatomically and functionally the medial pterygoid is the counterpart of the masseter, especially the superficial part, and vice versa (DuBrul, 1980).

The temporal

This is a bipennate muscle with fan-shaped origin and very large, tough, tendons of insertion into the coronoid process, deep temporal crest and anterior border of the ramus of the mandible.

The bulk and the length of the fibres are smaller than classically described, but longer than those of the medial pterygoid and masseter. Although it is the largest mandibular muscle it is not usually considered to be one of the power muscles attached to the mandible.

Origin

The major bulk of the temporal muscle arises from the temporal fossa of the lateral surface of the skull and the medial surface of the temporal fascia. It is divided into three areas, as viewed in lateral perspective, by the descriptive anatomists (Williams and Warwick, 1980).

1. The anterior temporal arises from that area of the temporal fossa, anterior to the craniomandibular joint. The fibres follow mainly a vertical course. The more anterior fibres may incline posteriorly towards their insertion, especially if the attachment extends well forwards into the frontal area (Figures 17 and 18). The anterior part of the temporal muscle arises off the temporal surface of the frontal and zygomatic bones down to the infraorbital fissure. It reaches its greatest thickness here. It also arises laterally from the deep surface of the temporal fascia and medially from the surface of the temporal fossa lateral to the infratemporal crest. The muscle is triangular in cross-section at this level (Figure 20).

   Some of the longest fascicles of the temporal muscle are found in this section, arising high on the anterior part of the temporal fossa and following a distally-inclined course to gain insertion in the temporal tendon (Figures 17 and 18). Lower down the lateral facing surface of the temporal fossa, short fibres arise which run with a lateral inclination to their insertion (Figure 19). Short fascicles also arise from the deep surface of the temporal fascia and follow a medially-inclined course to gain insertion in the temporal tendon (Figure 18). The temporal muscle is closely associated with the superior part of the lateral pterygoid muscle around the superior and lateral aspect of the infraorbital fissure and groove (Koritzer and Kenyon, 1983).
2. The middle temporal arises from the area approximately above the craniomandibular joint. These fibres course down with a slight anterior inclination (Figures 17 and 18).

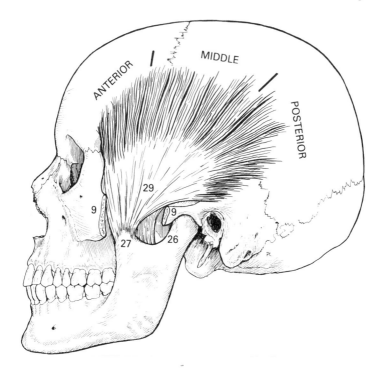

Figure 17 The temporal muscle in lateral perspective showing anterior middle and posterior parts, general alignment of muscle fasciculi and central tendon of insertion. The heavy black lines on the perimeter indicate the fibre origins and alignment with the mandible in the maximum intercusping position. The heavy lines indicate the fleshy part of the muscle. The less heavy lines indicate the central tendon. The zygomatic arch is removed to give visual access. The flexible, blade-like tendon of insertion is wide, antero-posteriorly, and narrow, medio-laterally. The tendon of insertion for the posterior horizontal fibres turns at less than 90 degrees to insert into the anterior half of the masseteric notch with the mandible in the maximum intercusping position: (9) cut zygomatic arch; (26) posterior edge of masseteric notch; (27) coronoid process; (29) central tendon of temporal muscle

3. The posterior temporal fibres arise posterior to the craniomandibular joint and run anteriorly, with increasing degrees of horizontal inclination. These fibres will operate with reduced efficiency due to the sharp change of direction of their tendon of insertion into the coronoid process of the mandible (Figures 17 and 18).

   When viewed in frontal perspective the muscle fibres gain insertion into the central, blade-like tendon from three sources of origin (Figure 19).

1. The temporal fossa on the side of the skull. The fibres have a lateral directional component from origin to insertion. Therefore they are capable of imparting a (small) medial vector to mandibular movement when they contract.
2. The deep surface of the temporal fascia. Only a relatively small portion arises from this structure. These fibres run almost vertically to insert into the central blade (Figure 19).

23

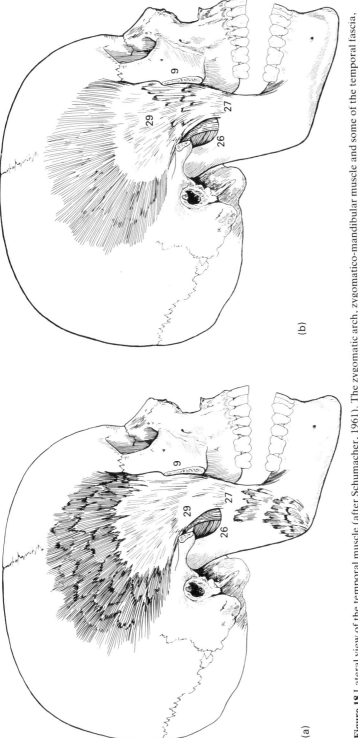

(a)

(b)

Figure 18 Lateral view of the temporal muscle (after Schumacher, 1961). The zygomatic arch, zygomatico-mandibular muscle and some of the temporal fascia, anteriorly, have been removed. (a) superficial and deep temporal fascia is indicated, together with general fibre alignment. The superficial part of the tendon of insertion is shown lower down. (b) Deep dissection of the fascia is indicated, together with general fibre alignment. The superficial part of the tendon of insertion is shown lower down. (b) Deep dissection of the temporal muscle indicates the fibre origin from the periosteum of the temporal fossa, general alignment and short course of the deep fibres, and the deep part of the central tendon extending high up into the muscle. Note the fibre origins deep, medially and inferiorly, in the anterior of the temporal fossa: (9) cut zygomatic arch; (26) posterior of masseteric notch; (27) coronoid process; (29) central tendon of temporal muscle

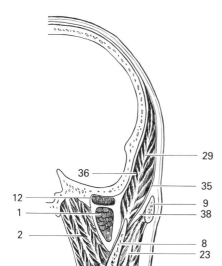

Figure 19 Frontal section through the skull, ramus, and the elevators and protruders of the mandible, to show the pennate nature of the elevator muscles. Note that the inferior fibres of the temporal muscle might impart a medio-lateral, horizontal component to the movement of the mandible (after Schumacher, 1961): (1) lower part of lateral pterygoid; (2) medial pterygoid; (8) cut ramus of mandible; (9) zygomatic arch; (12) superior part of lateral pterygoid; (23) masseter; (29) central, blade-like tendon of temporal muscle; (35) deep temporal fascia; (36) temporal muscle; (38) buccal pad of fat

3. The infratemporal surface of the skull just lateral to the infratemporal crest. Tendinous strips attached to the latter provide an origin for muscle fascicles. These fibres can likewise impart a slight medial vector (Zenker, 1954; Figure 19). Frontal section through the anterior part shows the deep temporal fascicles to be attached to a pointed medial off-shoot of the central tendon (Schumacher, 1961).

In horizontal section, at a level just above the zygomatic arch, the temporal muscle is seen to be triangular in cross-section with the broad base of the triangle running anteriorly towards the posterior surface of the maxilla (Figure 20). This is the thickest part of the muscle. From here the temporal muscle becomes thinner, superiorly and posteriorly, until, at the temporal line of the skull, it is barely perceptable.

Insertion

The temporal muscle is attached to the mandible by (Schumacher, 1961):

1. A large blade-like tendon of insertion into the coronoid process and down the anterior border of the ramus (Figures 17–19).

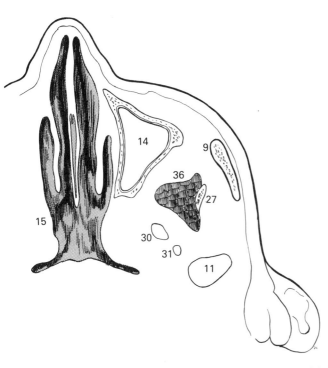

Figure 20 Diagrammatic representation of horizontal cross-section of skull, nose and maxilla at the level of the craniomandibular joint (after Prentiss, 1923) to indicate the shape and relative bulk of the temporal muscle in horizontal section. The outline of the coronoid process is dotted in. Note that temporal fasciculi arising off the frontal bone may impart a slight anterior component to the mandible (McCollum and Stuart, 1937) and that fasciculi arising antero-inferiorly off the lateral surface of the skull may impart a medial pull: (9) zygomatic arch; (11) condyle; (14) maxillary antrum; (15) nasal cavity; (27) outline of coronoid process; (30) foramen ovale; (31) foramen spinosum; (36) temporal muscle

2. In most cases, a separate tendon running down the deep temporal crest provides insertion for the fasciculi constituting the deep, medial and anterior part of the muscle. This is apparent only in frontal section at the level of the deep temporal crest.
3. ‘A small number of fasciculi insert directly into the medial surface of the ramus.

The central tendon of insertion, as seen in lateral perspective, reflects the overall outline of the temporal muscle. The lateral view of the temporalis shows considerable variation in size and outline from subject to subject, and on right and left sides in the same subject (Schumacher, 1961).

The blade-like tendon of insertion, while very strong, is flexible enough to bend when the mandible is depressed or the coronoid process is moved forwards, laterally or medially, and so avoids traumatizing the zygomatic arch of the maxilla (DuBrul, 1980).

According to origin and insertion the temporal muscle can be separated into two layers.

1. A thin superficial layer arising from the deep surface of the deep temporal fascia and inserting into the lateral surface of the central tendon. The fasciculi of this part are flattened down in a thin layer and only assume an oval bulk near their insertion (Figures 18 and 19).
2. The principal mass of muscle fibres arises from the side of the skull and the temporal surface of the frontal and zygomatic bones. The greatest bulk of the muscle is found anteriorly. The fasciculi of this layer form a thick fleshy part, anteriorly, and insert on the crest and medial surface of the central tendon or its medial off-shoots (Figures 18–20).

Temporal muscle insertion into the craniomandibular joint
The temporal muscle also arises off the inferior temporal surface of the skull lateral to the infratemporal crest, where it is closely associated with the lateral border of the superior part of the lateral pterygoid muscle and is inseparable from it. Some temporal muscle fasciculi arising from this area insert into the capsule–disc assembly of the craniomandibular joint over the articular eminence (Koritzer and Kenyon, 1983).

Posteriorly, temporal fasciculi often insert into the capsule and, through it, into the disc of the craniomandibular joint (Reese, 1954). Henke, (1863) classified this group of fasciculi as the 'minor temporal muscle'. Schumacher's dissections (1961), did not show well-defined attachments to the disc, but more widespread attachment to the anterior of the craniomandibular joint capsule.

The small muscle bundles from the temporal muscle which gain insertion into the capsule–disc assembly of the craniomandibular joint may play a role in the movements of the disc and condyle (Koritzer and Suarez, 1980).

Some biomechanical features

The fan-shaped nature of the temporal muscle origin, when viewed in lateral perspective, means that few fascicles will be aligned for maximum efficiency at any one time, i.e. directly aligned with the line of movement (Figures 17 and 18). Therefore, this muscle probably always will work at some mechanical disadvantage. The tendon from the horizontal fibres of the posterior part turns around the root of the zygomatic arch at an angle of 90 degrees or less. The acuteness of this angle will be emphasized when the mandible rotates down. Contraction of these fibres pulls the condyle up against the incline of the articular eminence (Figures 17 and 18) when the condyle–disc assembly is close to its fully seated position in the joint.

Its fan-shaped origin, bipennate structure and the architecture of the tendinous insertion means that, besides imparting a vertical pull on the mandible, there is some scope to develop vectors of movement in the sagittal and frontal planes. The posterior fibres in the main muscle mass can exert a posterior pull on the mandible when it is protruded.

The muscle is chiefly an elevator and retractor of the mandible and if the muscle is activated sequentially, from anterior to posterior, the direction of pull of the contracting fibres will be the same as the upward swing of the coronoid process as the mandible is elevated from a depressed position (Kraus, Jordan and Abrams, 1969).

The power of the muscle is considerably increased by its bipennate architecture and the origin of fibres off the medial surface of the temporal fascia, as well as off the temporal fossa.

Despite the relatively short length of the fibres and long tendinous insertion, there is enough inherent elasticity to allow easy stretching to accommodate the full range of mandibular border movement.

The posterior, horizontal fibres of the temporal muscle insert into a tendon which turns over the edge of the root of the zygomatic process of the temporal bone. Its insertion is into the deepest part of the concavity of the masseteric notch. This, in turn, is distal to the point at which the tendon turns over the zygomatic process (Figure 17). Hence contraction of the posterior fibres, with the condyle almost or fully seated in the glenoid fossa, will draw the condyle–disc assembly into close contact with the postero-superior aspect of the articular eminence (DuBrul, 1980).

In a protruded position, the posterior part of the tendon of insertion is pulled anteriorly and inclined such that it is approximately parallel with the slope of the articular eminence. Contraction of the posterior part with the mandible in this position will retrude the condyle–disc assembly (Figure 18).

The middle temporal fibres have an oblique down and forward inclination and exert a definite retrusive component (Figures 17 and 18).

With its bipennate, fan-shaped, muscle fibre alignment the temporal muscle is well equipped to contribute to the final, fine adjustment of mandibular position – antero-posteriorly, medio-laterally and vertically.

Electromyographic studies show the temporal muscle to be active (Møller, 1966):

1. In all elevation strokes.
2. In retrusion of the mandible, especially the posterior temporal fibres.
3. In the final positioning of the mandible into the maximum intercusping position.
4. In maintaining elevation of the mandible during gliding, tooth contact movements, especially when the posterior teeth are separated in such movements (Williamson and Brandt, 1983).

Nerve and blood supply

The temporal branches of the mandibular division of the fifth cranial nerve carry innervation for the temporal muscle. The blood supply is derived from the temporal branches of the auriculo-temporal artery.

Summary of the anatomy of the temporal muscle

1. It is superficially placed on the side of the skull.
2. It is the largest of the mandibular muscles.
3. It has a very widespread and diverse area of attachment at its origin.
4. Its attachment to the mandible is concentrated into a relatively small area.
5. The mechanism used to achieve attachment to the mandible is a large, flattened and flexible tendon of insertion whose outline broadly conforms to the outline of the area of origin.
6. By far the greatest mass of the muscle is located anteriorly in the retro-orbital part of the temporal fossa.
7. In horizontal section, at the level of the zygomatic arch, the muscle is triangular in cross-section with its broad base towards the temporal surface of the zygomatic bone.

8. The thickness of the muscle decreases postero-superiorly to become almost imperceptible at the temporal line.
9. The three-dimensional alignment of its muscle fibres enables it to move the mandible through the three planes of space.
10. It is clearly associated with its neighbouring muscles.
11. Fasciculi arising around the craniomandibular joint are attached to the capsule or disc of the joint and may influence positioning of the disc.
12. The temporal fascia is split and filled with fat above the zygomatic arch. The deep layer of the temporal fascia is tough and aponeurotic. The latter feature may prevent satisfactory palpation of the muscle tissue deep to it.

The mandibular capsular muscle

This is a small muscle slip arising from the edge and related areas of the posterior half of the masseteric notch of the mandible. The fibres take a posterior, superior and medial course to insert into the lateral and anterior parts of the capsule and disc of the craniomandibular joint (Koritzer and Suarez, 1980; Figure 21).

This small muscle mass is quite discrete. It is neither part of the masseteric complex, nor is it associated with the temporal muscle. Both masseter and temporal muscle are mainly tendinous at this level (Koritzer and Suarez, 1980).

It is uncommon to find a muscle that originates on moving bone and inserts into a moving disc.

The function of the mandibular capsular muscle is probably complex. Functions postulated by Koritzer are:

1. To provide a measure of balance by opposing the superior part of the lateral pterygoid. This function could be carried out indirectly, by means of sensory input to the central nervous system initiated by tension in the capsular muscle

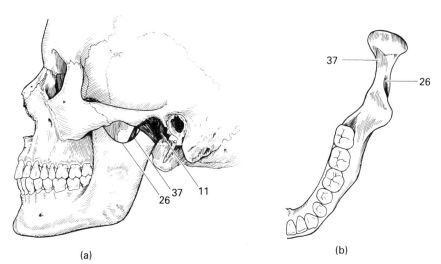

(a)

(b)

Figure 21 The mandibular capsular muscle: (a) as seen in sagittal perspective; (b) as seen from above; (11) craniomandibular joint capsule; (26) edge of masseteric notch; (37) mandibular capsular muscle

tissue. The physical bulk of the mandibular capsular muscle is only one-fifth of that of the upper part of the lateral pterygoid (Koritzer and Suarez, 1980), so any action it has in balancing the pull of the upper head of lateral pterygoid is unlikely to be due to the physical strength of 'mass action' of the capsular muscle.

2. By contracting, it could help to reduce torsion in the disc over the head of the condyle.
3. It can apply some lateral pull to offset the medial component of force applied to the disc by the lateral pterygoid.
4. It can help control retrusion of the disc supero-posteriorly, along the articular eminence, by releasing tension as this occurs, even though its own origin is moving in the same direction.

The zygomatico-mandibular mass of muscle tissue

The temporal and masseter muscles are only separable, anteriorly, where the buccal pad of fat extends between them (Figure 22). The deep surface of the deep part of the masseter, anterior to the tubercle of the zygomatic arch, and the temporal muscle overlain by it, intermingle and cannot be blunt-dissected apart

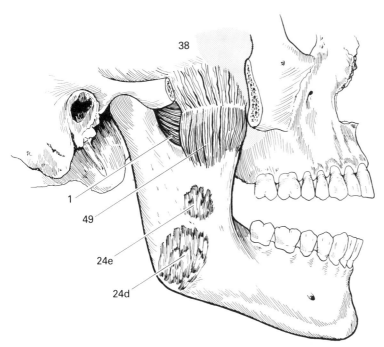

Figure 22 Lateral view of zygomatico-mandibular muscle (49) (after Schumacher, 1961). The zygomatic arch has been cut away for clarity. Fibres arise from the deep surface of the zygomatic arch and deep temporal fascia and insert into the central tendon and lateral surface of the ramus and the coronoid process of mandible: (1) lateral pterygoid muscle; (24e) upper mandibular septum (masseter); (38) buccal pad of fat; (49) zygomatico-mandibular muscle

over a distance of about 3 cm. This part of the mandibular muscle tissue is named the zygomatico-mandibular muscle. It is not circumscribed by any perimuscular planes (Schumacher, 1961; DuBrul, 1980; Koritzer and Suarez, 1980).

Some zygomatico-mandibular fasciculi insert into the capsule–meniscus assembly of the craniomandibular joint (Rees, 1954; Schumacher, 1961; Koritzer and Suarez, 1980; Widmalm, Lillie and Ash, 1987).

The lateral pterygoid muscle

Introduction

The lateral pterygoid muscle occupies a deep and hidden position (Figures 23 and 24). It was ascribed the function of protracting the mandible. Because mandibular movement was a relatively small problem in former times, and because of its deep position and small size, little attention was given to this muscle mass and it was treated almost dismissively by the classical anatomists.

With the development of refined maxillo-facial surgical techniques and the deepening understanding and importance of functional and dysfunctional aspects of

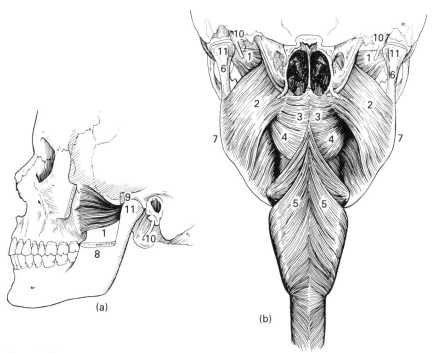

Figure 23 Diagram to show the deep position of the lateral pterygoid muscle on the base of the skull posterior to the maxilla (after Prentiss, 1923): (1) lateral pterygoid muscle; (2) medial pterygoid muscle; (3) superior constrictor muscle; (4) levator palati muscle; (5) middle constrictor muscle; (6) posterior border of ramus; (7) inferior border of mandible; (8) cut ramus of mandible; (9) cut zygomatic arch; (10) styloid process; (11) condyle; (a) sagittal section viewed from the lateral side; (b) paracoronal section viewed from the postero-inferior aspect to show the position of the pterygoid muscles in the side of the nasopharynx

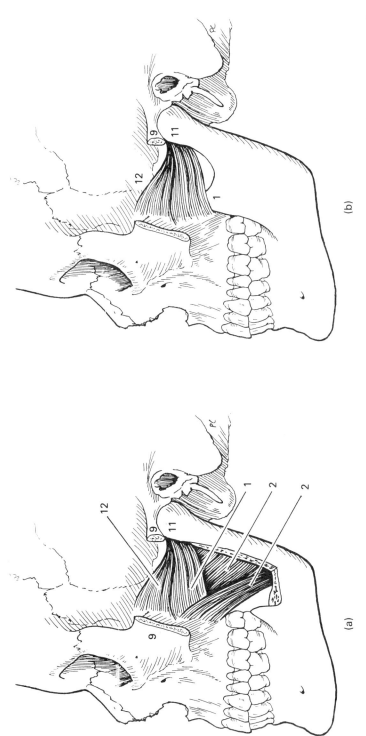

(a)

(b)

Figure 24 Lateral pterygoid from the lateral perspective, with the zygomatic arch, temporal and masseter muscles and ramus of mandible cut away to expose it. The two parts are divergent and separate clearly when viewed anteriorly, but posteriorly they are less distinctly separate and appear to fuse just before their insertion into the disc (mainly upper head) and fovea (mainly lower head). There is a wide area of origin and a narrow insertion. This gives rise to different vectors when the different parts contract. Some authorities consider that the two parts are quite separate and are enclosed within their own fascial sheaths with a fibrous layer between them (Honeé, 1972). On close inspection, however, the fibres of the two parts are seen to be aligned differently. (a) After Prentiss, 1923: (1) lower part of lateral pterygoid; (2) medial pterygoid (upper portion of superficial head removed); (9) cut zygomatic arch; (11) condyle of the craniomandibular joint; (12) superior part of the lateral pterygoid. (b) Uppermost fibres of the inferior head are left out to emphasize the separateness of the two heads anteriorly, the angular relationship of the upper head with the horizontal plane and the wide area of origin and narrow area of insertion that may enable variation in the line of traction

mandibular movement there has been extensive investigation of the anatomy and physiology of this muscle over the past 30 years. Because the significance of this work for everyday applied dentistry is not widely appreciated, and because of the scattered nature of the source of the material, the following rather extensive review is offered.

The lateral pterygoid muscle

The lateral pterygoid muscle is situated deep to the ramus of the mandible and the temporal muscle in the side wall of the nasopharynx. It lies just below the base of the skull, posterior to the maxilla and anterior to the posterior border of the ramus of the mandible (Figures 23 and 25).

Owing to the hidden position of this tissue its function had to be postulated and assessed on the basis of mechanical principles applied to knowledge of the attachments of the muscle.

The blood supply of the lateral pterygoid is derived from the intimately related maxillary artery and the venous plexus associated with it (Gray, 1858). It is innervated by the lateral pterygoid branches of the mandibular division of the fifth cranial nerve (Gray, 1858).

Two parts of this muscle are described, apparently on the basis that there is fusion of both parts, posteriorly, just in front of the insertion.

Most modern investigators who have dissected carefully have shown that:

1. There is great variability in the interdigitative union of the muscle fibres of both parts (Prentiss, 1923; Christensen, 1969; Porter, 1970; Honèe, 1972; Mahan, 1980; Sicher, 1960; Schumacher, 1961; Widmalm, Lillie and Ash, 1987).
2. Superior and inferior parts have been shown to have their own fascial sheaths. Distally where the muscles 'meet' they may still be separable by a thin fascia, (Honèe, 1972).

Owing to this and various other anthropological and developmental factors many of the above authorities regard the two 'heads' of the lateral pterygoid as quite separate and distinct entities (Prentiss, 1923; Christensen, 1969; Koritzer, 1983). Much functional evidence also supports this view (McNamara, 1973; Lipke et al., 1977, Koole et al., 1984; Widmalm, Lillie and Ash, 1987).

Superior part ('head') of the lateral pterygoid

Origin
The superior head of the lateral pterygoid originates from the infratemporal surface of the great wing of the sphenoid bone. It extends as far forwards as the inferior orbital fissure (Figures 25–28) and the muscle tissue may be continuous throughout the orbital fissure with the orbitalis muscle (Koritzer, 1983). Laterally, the origin extends to the infratemporal crest where it blends imperceptibly with the temporal muscle (Schumacher, 1961; Christensen 1969; Koritzer, 1983). The muscle fibres are tightly integral with the underlying periosteum up to a point about 10 mm anterior to their insertion.

Insertion
The muscle inserts itself into the capsule and disc of the craniomandibular joint. Upper head fibres blend with those of the lower head to gain insertion into the neck

of the condylar process of the mandible. (Schumacher, 1961; Christensen, 1969; DuBrul, 1980; Williams and Warwick, 1980; Widmalm, Lillie and Ash, 1987).

While the actual point of insertion of the superior part of the lateral pterygoid has been the subject of controversy, most of the investigators of recent times have described the insertion as being chiefly into the disc and capsule (Figures 25, 26 and 29) whenever the superior and inferior parts of the muscle are separated from each other, (Prentiss, 1923; Christensen, 1969; Porter, 1970; Honèe, 1972; Mahan, 1980). Most of the modern authors further emphasize the insertion of this muscle into the medial side of the disc (Neff and Suarez, 1983; Moffett, 1984). There is evidence of a mingling of superior and inferior part fibres and of a criss-crossing of fibres from the upper and lower parts such that some upper part fibres attach to the fovea of the condyle neck and some lower head fibres attach to the capsule or disc (Sicher, 1951; Schumacher, 1961; Mahon, 1980; Widmalm, Lillie and Ash, 1987).

Moffett (1984) points to the embryological evidence that clearly establishes the origin of the disc of the craniomandibular joint as part of the lateral pterygoid muscle which becomes intercepted by the developing parts of the craniomandibular joint. He emphasizes the similarities with the development of the knee and clavicular joints. Moffett has shown that the intercepted lateral pterygoid tendon becomes the medial part of the craniomandibular joint disc and that, antero-medially, the disc clearly merges into muscle fibres of the superior part of the lateral pterygoid muscle.

Since the disc is firmly attached to the condyle at the lateral and medial poles (Ärstad, 1954) and, posteriorly, to the neck of the condyle between the poles (Reese, 1954), the superior head of the lateral pterygoid may be considered to have a stirrup-like ligamentous attachment to the condylar process of the mandible (Juniper, 1981).

Thus, although the traction applied directly by the superior head of the lateral pterygoid is mainly to the disc, it will also be applied indirectly to the condyle through the disc as long as the attachments of the disc to the condyle are properly maintained. If the attachment of the disc to the condyle is disrupted, then this muscle will tend to displace the disc of the condyle in a medial and anterior direction (Mahan, 1980). The superior part, through the fibres that interdigitate with the inferior part and pass down to the pterygoid fovea, may have a direct, small influence on the mandible (Widmalm, Lillie and Ash, 1987).

The superior part is usually described as being 4–5 cm long (Schumacher, 1961; Honèe, 1972). Koritzer (1983) found it to be somewhat longer, 5–6 cm. It is about 1 cm wide and less than 0.5 cm thick. The fibres course downwards, backwards and laterally to gain insertion into the capsule or disc of the joint (Figures 24–27). Thus it is a quadrilateral strap which 'lies on its back' almost in the horizontal plane (Figures 25 and 27). Its greatest cross-section is about 0.5–1 cm distal from the back of the maxilla (Widmalm, Lillie and Ash, 1987).

Its superior aspect is tightly integral with the periosteum and its continuous origin can be elevated with a periosteal elevator. There are no limiting fascial planes or compartments (Koritzer, 1983). The periosteum is freely continuous through the foramina ovale and spinosum with the dura mater of the middle cranial fossa (Koritzer, 1983; Figure 27).

The length of the upper part is commensurate with significant function. Its attachment to the medial aspect of the disc is very firm and it is very closely applied to the periosteum of the infratemporal fossa. There is little freedom between the bony and the disc attachments since the distance between them measures

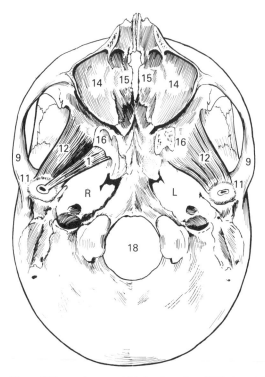

Figure 25 Lateral pterygoid (after Prentiss, 1923) viewed from below: (R) right side sectioned to show superior (12) and inferior (1) parts; (L) left side showing superior part only (note attachment to capsule–disc assembly on the medial side); (1) inferior head of lateral pterygoid; (9) zygomatic arch; (11) condyle – capsule of craniomandibular joint; (12) superior head of lateral pterygoid; (14) maxillary antrum; (15) superior meatus of nose; (16) bony origin of medial and lateral pterygoid plates (cut off); (17) roof of nasopharynx; (18) foramen magnum

approximately 1 cm (Figures 25 and 29). This would appear to imply an isometric type of activity for the upper part of the muscle, (Koritzer, 1983).

The origin of the superior part of the lateral pterygoid on the roof of the infratemporal fossa, is closely related, medially, to the mandibular division of the fifth cranial nerve at the foramen ovale and to the middle meningial artery at the foramen spinosum (Williams and Warwick, 1980; Koritzer, 1983).

Laterally, at the infratemporal crest of the great wing of the sphenoid, the superior head of the lateral pterygoid blends with the temporal muscle (Schumacher, 1961; Christensen, 1969; Koritzer, 1983; Widmalm, Lillie and Ash, 1987).

Anteriorly, the superior part of the lateral pterygoid muscle is continuous with the anterior part of the temporal muscle along the superior and lateral borders of the inferior orbital fissure (Schumacher, 1961; Koritzer, 1983) and may blend with fibres of the orbitalis muscle.

The superior part of the lateral pterygoid is a thin strap of muscle tissue. The inferior part of the muscle has a relatively thick belly which can generate two-and-a-half times more power and is three times heavier than the superior part (Schumacher, 1961).

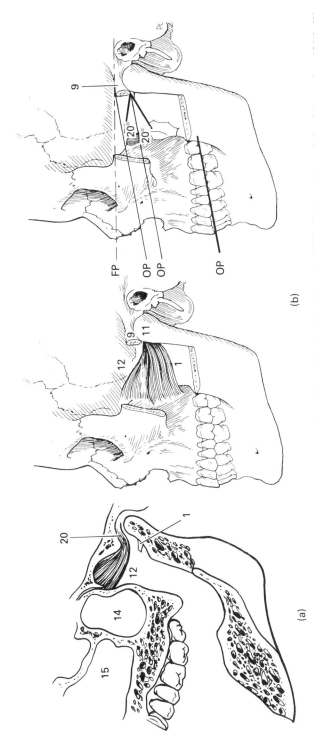

(a)

(b)

Figure 26 Superior part of the lateral pterygoid muscle, in sagittal view: (a) medial view (after Prentiss, 1923); (b) lateral views (after Schumacher, 1961); (1) tendon of inferior head of lateral pterygoid; (12) superior head of lateral pterygoid; (14) maxillary antrum; (15) superior meatus of nose; (20) anterior band of the craniomandibular joint; (FP) Frankfort plane; (OP) occlusal (Camper's) plane and planes parallel to it above and below the joint

The muscle fibres of the superior part are aligned to exert a force vector antero-medially to the sagittal plane, at an angle of 26 degrees (Honèe, 1972; Figures 24, 26 and 28). The superior head fibres are aligned to pull vertically, at an angle of 12 degrees to the Frankfort plane (Lord and Hanover, 1937; Honèe, 1972; Figures 24 and 26).

Function of the superior head of the lateral pterygoid

The superior part of the lateral pterygoid muscle is active during retrusion of the ipsilateral condyle. Its function is to guide and stabilize the condyle–disc assembly against the articular eminence during this phase of translatory movement (Vaughan, 1955; McNamara, 1973; Lipke *et al.*, 1977; Bell, 1982; Koole *et al.*, 1984; Widmalm, Lillie and Ash, 1987), and possibly also in the postural position (Juniper, 1981). The superior part of the lateral pterygoid is also active during clenching of the jaw (Widmalm, Lillie and Ash, 1987).

Owing to its small bulk and isometric activity this muscle may be more prone to dysfunctional involvement (Koole *et al.*, 1984), than other muscle tissue in the masticatory system.

The inferior part ('head') of the lateral pterygoid

Origin

The inferior 'head' of the lateral pterygoid originates from the lateral surface of the lateral pterygoid plate and back of the maxilla up to the infraorbital fissure

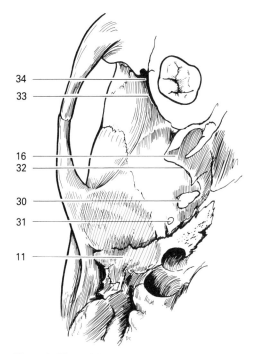

Figure 27 View of the skull to show some relationships of the superior part of the lateral pterygoid muscle: (11) condyle – capsule of craniomandibular joint; (16) pterygoid fossa; (30) foramen ovale; (31) foramen spinosum; (32) pterygo-maxillary fissure; (33) infratemporal crest; (34) inferior orbital fissure

(Figures, 23, 24, 28 and 29). The lateral pterygoid plate is situated to the back of the maxilla, is aligned vertically and angled postero-laterally (Figures 27–29). The plate measures approximately 30 mm in height and 10 mm in width.

Insertion
The muscle inserts itself into the fovea of the anterior surface of the neck of the condyle process of the mandible. This is a small area measuring approximately 10 mm × 5 mm (Figures 25, 26 and 29).

The fibres constitute a belly-shaped mass which runs postero-laterally, at an angle of 30–45 degrees to the sagittal plane, from the extensive fan-shaped origin to the small area of insertion (Figures 23, 24, 28 and 29). This gives the muscle a triangular shape in lateral perspective. The greatest width of the muscle is in the vertical plane, along its origin. The smallest dimension of the muscle is its depth in the lateral and medial dimensions (Figures 28 and 29). Most of the fibres have an upward alignment to their insertion of about 20 degrees (Figure 30) in sagittal perspective. The inferior head of the lateral pterygoid is flat in the vertical plane and has a much longer tract of fleshy muscle between origin and insertion than the upper head. This arrangement would appear to favour isotonic muscle activity. The internal structure of the inferior 'head' of the lateral pterygoid muscle shows some fibrous laminae (septa) and a pennate arrangement (Schumacher, 1961; Widmalm,

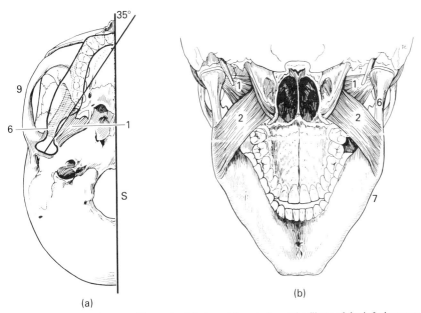

(a)   (b)

Figure 28 The lateral pterygoid muscle. (a) viewed from below. The fibres of the inferior part run from origin to insertion in the postero-lateral direction making an angle of 35 degrees with the sagittal plane: (1) lower part of lateral pterygoid. (b) The pterygoid from the postero-inferior aspect. The medio-lateral inclination of the fibres of the inferior head of the lateral pterygoid make an angle of 35 degrees with the sagittal plane. In this view note that the medial pterygoid (2) masks digital access to the lateral pterygoid (1) from the pharynx: (1) lateral pterygoid; (2); medial pterygoid; (6) posterior border of ramus; (7) inferior border of mandible; (9) zygomatic arch; (S) mid-sagittal plane

Lillie and Ash, 1987). The septa are situated mainly at the condylar end and gain insertion into the neck of the condyle. There are septa also at the origin on the lateral pterygoid plate.

Function of the inferior part of the lateral pterygoid
The functions of the inferior part of the lateral pterygoid are to:

1. Protrude the ipsilateral condyle–disc assembly in order to manoeuvre the hinging axis of the mandible into functional positions (Ferrein, 1774; Ekholm and Sirrila, 1960).
2. Rotate the mandible around the hinge axis (Grant, 1973a) in simple depression.
3. Stabilize the protruded ipsilateral condyle on the articular eminence (Vaughan, 1955; Sicher 1960; Dawson, 1974).
4. Prevent the condyle being displaced superiorly and distally (DuBrul, 1980; Kawamura, 1968; Widmalm, Lillie and Ash, 1987).

The inferior head of the lateral pterygoid is a bulky, short and powerful muscle. Its chief function is to produce translation of the ipsilateral condyle–disc assembly in the superior compartment of the craniomandibular joint. Simultaneous action of the right and left muscles will cause symmetrical, bilateral protrusion of the mandible and prevent excessive distal and superior displacement of the condyle.

Unilateral action of this muscle will cause a lateral swing of the mandible towards the opposite side (Figures 60 and 62). The effect is due to the oblique alignment of the fibres between origin and insertion in the horizontal plane (Figures 28 and 29) they run at a 30–45 degree angle to the sagittal plane. The inferior part of the lateral pterygoid is involved in all functional movement of the mandible and in the maintenance of the postural position (Juniper, 1981).

Relationship of the two parts of the lateral pterygoid

The width of the inferior part of the lateral pterygoid is maximum vertically – the muscle 'stands on its side' in the vertical plane running from origin to insertion (Figure 24). The superior part, in general, lies 'flat on its back' towards the horizontal plane (Figures 25–27). The greatest widths of the two muscles are at right angles to each other.

Viewed in the sagittal perspective, the long axes of the general alignment of the fibres of the upper and lower parts of the lateral pterygoid muscle diverge from each other by about 30–40 degrees, anteriorly, and are related closely near to their insertion (Figure 26; Schumacher, 1961).

Like the temporal muscle, the origin of the lateral pterygoid spreads in a fan-shaped fashion when viewed in sagittal perspective (Figure 24). The muscle fasiciculi then converge to insert themselves into the pterygoid fovea and capsule–disc assembly of the craniomandibular joint. Fibrous laminae in the muscle mass increase the scope for attachment of muscle fibres and aid insertion to the constricted area below the craniomandibular joint.

This structural distribution enables the muscle mass to act as a stabilizer for the joint, or as a protractor or depressor of the jaw, or both the latter simultaneously.

The bulk and power of the inferior part of the muscle is much greater than that of the superior part (Schumacher, 1961; Honèe, 1972). The inferior part has the potential for full isotontic contraction while the superior part is probably limited in

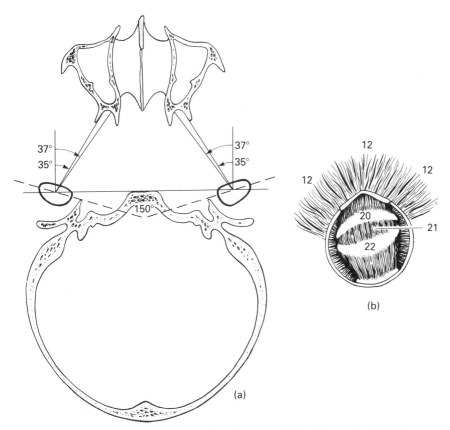

Figure 29 (a) Horizontal perspective of the lateral pterygoid (after Schumacher, 1961) showing that the inferior part makes an angle of 33 degrees, and the superior part an angle of 37 degrees, with the sagittal plane: (b) superior head from above, inserted into the meniscus of the craniomandibular joint over its antero-medial two-thirds; (12) fibres of the superior part of the lateral pterygoid; (20) anterior band of meniscus; (21) intermediate zone of meniscus; (22) posterior band of meniscus

this respect because of its attachment over most of its length to the periosteum (Koritzer, 1983).

Owing to the main fibre alignment, the vector of the pull exerted by the inferior 'head' probably will be different to that of the superior in the horizontal plane (Figures 25, 28 and 29; Prentiss, 1923; Honèe, 1972). The lower head will pull slightly downwards in the vertical plane (Figure 24). The superior part, which is aligned upwards in the vertical plane at an angle in the range 10–20 degrees to the Frankfort plane, will pull the disc with an antero-medial and inferior vector (Figures 24 and 26). The insertion of the superior part passes over the crest of the articular eminence to insert into the disc and will tend to hold the condyle–disc assembly in contact with the articular eminence (Figure 26; Schumacher, 1961; Honèe, 1972; Bell, 1982).

The functions of the two parts are quite different, but both are concerned with manoeuvring and stabilizing the ipsilateral condyle–disc assembly.

Electromyography of the lateral pterygoid

Good electromyographic studies of the lateral pterygoid are rare:

1. Until recently the trend was to consider that the superior and inferior parts of the muscle had the same function (Gray, 1858; Cunningham, 1902; Morris, 1953). Therefore, no attempt was made to put electrodes into both superior and inferior parts.
2. Placement of electrodes in the muscle mass without causing pain, haematoma and interference with function was difficult (McNamara, 1973). One might question whether earlier investigators were sure that the electrodes were placed in the lateral pterygoid muscle tissue at all. In recent times, with better definition of anatomical features and very careful technique, it has been possible to locate fine wire electrodes in the two parts of the muscle (McNamara, 1973; Koole *et al.*, 1984; Widmalm, Lillie and Ash, 1987). Because of these difficulties higher primate models closely resembling man have been used (McNamara, 1973).

These more recent investigations have found the inferior part of the lateral pterygoid to be active in protraction and depression of the mandible in combination with the suprahyoid muscles (Lipke *et al.*, 1977; Widmalm, Lillie and Ash, 1987). Widmalm and his co-workers found the inferior part to be very active in voluntary tooth gnashing. They noted that distal manipulation of the mandible caused marked activity of the lower head.

The superior head was found to be active during mandibular retrusion in conjunction with the mandibular elevators. This would support the suggestion that the superior head probably supplies a stabilizing, guiding action for the craniomandibular joints in translatory movement. Widmalm, Lillie and Ash (1987) noted considerable activity of the superior part during tooth clenching.

Lipke *et al.* (1977), in an electromyographic study of the lateral pterygoid function in man, confirmed McNamara's findings. They demonstrated that the superior part was active in mandibular elevation, in synergy with the elevators, and silent during mandibular protrusion or depression when the suprahyoid group is active.

Juniper (1981), using human subjects, confirmed McNamara's and Lipke's findings yet again. In addition, he demonstrated activity of the superior part of the lateral pterygoid when the mandible was in the postural position. Koole *et al.* (1984) report activity of the inferior part of the lateral pterygoid during mandibular protrustion and depression in healthy human masticatory systems. During mandibular protrustion the superior part was silent. The reverse situation occurred during mandibular retrusion. Here, the superior part was found to be progressively more active as the condyle moved postero-superiorly and the inferior part was inactive.

Other electromyographic studies have noted the activity of the superior part of the lateral pterygoid (Carlsöö, 1956; Hickey, Stacey and Rinear, 1957; Griffin and Munro, 1969). McNamara (1973) strongly favours a role for the superior part in positioning and holding the condyle–disc assembly on the articular eminence during translations in the craniomandibular joint. McNamara also points out that firm apposition of the condyle–disc assembly on the eminnence may not be essential during mandibular depression.

Ekholm and Siirila (1960) are quoted by McNamara as finding, in an electromographic study, that the activity of the lateral pterygoid was not essential for mandibular depression, but that the muscle was active during mandibular depression as part of habitual movements.

Considerable disturbance of normal electromyographic activity of the lateral pterygoid muscle was found in patients suffering dysfunction of mandibular movement. The normal patterns of activity were not evident and there was prolongation of the normal silent periods, (Koole *et al.*, 1984).

Clinical studies of the lateral pterygoid

In a longitudinal clinical study of the effects of paralysis of the lateral pterygoid the following features were documented (Vaughan, 1955).

1. An inability to protrude the mandible on the paralysed side.
2. An inability to carry out gliding, tooth contact movements in an antero-lateral direction on either side when only one pterygoid was paralysed.
3. The normal chewing cycle was not possible.
4. On mandibular depression the mandible is displaced to the affected side when one lateral pterygoid is paralysed.
5. Radiographic evidence of distal displacement of the condyle–disc assembly in the craniomandibular joint on the affected side.
6. Vaughan found that the activity of the paralysed lateral pterygoid muscle could not be compensated for by the other mandibular muscles. He felt that the lateral pterygoid was continually active, to the extent that it drew the condyle–disc assembly anteriorly, to a degree sufficient to maintain steady contact between the assembly and the articular eminence, both at rest and during functional movements of the mandible.

Summary

1. The descriptive anatomy of this muscle tissue is now accurately defined.
2. The probable functional roles of the muscle tissue in the upper and lower parts of this muscle are shown to be different by electromyographic investigation.
3. Anatomically and functionally, the upper and lower parts of the lateral pterygoid are probably two different muscles.
4. In general, the role of both parts is concerned with the positioning and stabilization of the condyle–disc assembly on the articular eminence during the performance of functional movements and probably also in the postural position. Specifically, the superior part is active only when the ipsilateral condyle–disc assembly is being retruded toward the glenoid fossa; and during clenching. Likewise, the inferior part is active when the condyle–disc assembly is being protruded and stabilized in a protruded position. Hence, the lateral pterygoid is involved in all mandibular movements.
5. The normal activity of both the superior and inferior parts of the lateral pterygoid is considerably disturbed when dysfunction of mandibular movement or internal derangement of the craniomandibular joints is present.
6. Severe interference with normal function of the lateral pterygoid results in considerable limitation or failure of mandibular function. Paralysis of one lateral pterygoid muscle can not be compensated for adequately by the remaining muscles which move the mandible.

Biomechanics of the lateral pterygoid

Accurate delineation of the attachments of the muscle have allowed data on the following physical factors to be determined.

The overall length of the lateral pterygoid is 40–50 mm (Schumacher, 1961; Honèe, 1972; Koritzer, 1983).

The internal structure of both parts of the muscle shows that there is a regular, near-parallel relationship between the muscle fibres and that low pinnation angles exist on the intramuscular fibrous septae in the inferior part (Schumacher, 1961; Honèe, 1972; Grant, 1973a; Widmalm, Lillie and Ash, 1987).

The average length of the muscle fibres in both parts is approximately 25 mm (Schumacher, 1961; Honèe, 1972). Hence the inferior part could change its length by about 10–15 mm (Honèe, 1972). The superior part, having a thickness of only 5–10 mm and being closely attached to the underlying periosteum along its length (Koritzer, 1983), would have limited scope for shortening.

The wet weight of the inferior part (approximately 3 g) is three times greater than that of the superior part (Schumacher, 1961; Honèe, 1972).

The cross-section of the inferior head is about 10 mm^2 and the superior head about 4 mm^2 (Honèe, 1972).

Honèe's investigation broadly confirms that of Schumacher (1961). Investigators differ in the average angulations they give for the average alignment of the muscle fasciculi of the superior and inferior parts in the vertical and horizontal planes. This

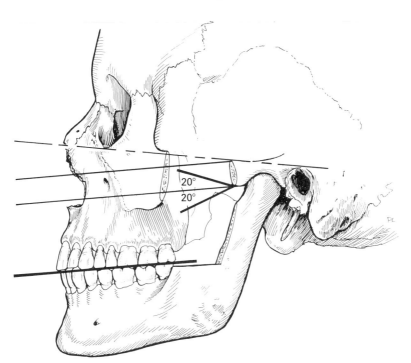

Figure 30 Result of alignment of superior and inferior parts of the lateral pterygoid muscle. The two parts diverge, anteriorly, at an angle of 40 degrees. The superior head makes an angle of 12 degrees (approximately) with the Frankfort plane. The inferior head makes an angle of 28 degrees (approximately), with the Frankfort plane open in the opposite direction (after Schumacher, 1961)

is to be expected in view of the varying morphology from cadaver to cadaver, the lack of standardized procedure for the measurements in different investigations and the small numbers involved because of the difficult and prolonged nature of the work. Nevertheless, definitive trends are clearly established and differences of detail are unlikely to be significant for understanding the functions of the muscle tissue.

The following conclusions can be drawn:

1. In lateral perspective, the superior part fasciculi are aligned downwards from origin to insertion. The angulation varies from 10 to 25 degrees to the Frankfort plane with the mandible in the postural position. The tendon of insertion passes over the crest of the articular eminence. The inferior head fasciculi arise from a fan-shaped origin and present an average alignment which is upwards from origin to insertion, approximately 10–30 degrees to the Frankfort plane (Figure 30).
2. In horizontal perspective, both parts make an angle with the sagittal plane since they run laterally from their origin (Figure 29). Some authorities give the largest angulation in this plane to the inferior part (e.g. Prentiss, 1923; Honèe, 1972; Juniper, 1982) while others find that the superior part makes a bigger angle with the mid-sagittal plane (e.g. Schumacher, 1961; Koritzer and Suarez, 1980). Examination of a large number of skulls leads the author to favour the former concept with only a small angular difference between the alignment of the two parts. The vector of force of the inferior part is directed medially at an angle which varies but is in the range of 30–45 degrees to the sagittal plane through the craniomandibular joint (Figures 24 and 27–29).

Theoretical and experimental exposition of the biomechanical effects produced by the lateral pterygoid

With clarification of the muscle attachments and the direction of pull of the main groups of muscle fibre, the axes about which the mandible moves when muscular pull is exerted become important. When the assumption is made that rotatory motion of the mandible in the sagittal plane occurs around a stationary axis, then the conclusion may be drawn that the lateral pterygoid cannot exert a sufficient turning moment to make an effective contribution to the depression of the mandible. On this basis it may be said to be only a protractor of the mandible (Carlsöö, 1956; Woelfel, Hickey and Rinear, 1957). However, the mandible is rarely rotated around a stationary axis alone in habitual function (Posselt, 1952). Thus, if the activity of the lateral pterygoid is considered in relation to the instantaneous axes of rotation that develop in ordinary functional movements (Luce, 1889; Ulrich, 1896; Nevakari, 1956), then turning moments which are effective in achieving rotations are quite feasible (Grant, 1973a).

Others have pointed out that the lateral pterygoid probably has a role to play in positioning and stabilizing the condyle–disc assembly in mandibular function (Vaughan, 1955; Sicher, 1960; Christensen, 1969; Bell, 1982; Widmalm, Lillie and Ash, 1987).

The effects produced by contraction of the lateral pterygoids were simulated in cadavers using correctly aligned string to move the mandible (Ulrich, 1896). Protraction of the condyle–disc assembly was thereby demonstrated. The need for the coordinated action of many muscles to produce functional-like movements become very apparent in this experiment. Contraction of the lateral pterygoids was shown to reduce the necessity for extreme shortening of the digastrics to achieve mandibular depression. Findlay (1964) demonstrated similar effects in the cadaver and dry skull. Lord and Hanover (1937), using an elaborately mounted dry skull with artificially constructed discs in the joints and a combination of springs and pulley cords to simulate muscle action, also demonstrated similar effects.

Grant (1973b) analysed the likely effects of lateral pterygoid activity by superimposing instantaneous centres of rotation, as determined by functional and habitual movements, on a skull in side view. In this way he demonstrated that the inferior head could exert a moment of force which would depress the mandible from the postural position and which increased in degree as the mandible was depressed. Furthermore, he showed that the superior head of the lateral pterygoid could exert a small elevating moment from a depressed mandibular position, but could exert neither an elevating nor a depressing moment when the jaw was in the postural position. He therefore concluded that the superior and inferior parts of the lateral pterygoid muscle applied force in different directions and, mechanically, had antagonistic actions. He emphasized that the lateral pterygoid could protract the mandible only with the aid of the elevators, to prevent rotations.

Grant's investigations leave the role of the superior part of the lateral pterygoid unclear. While it appears that the superior part could assist elevation of the mandible, he favours the view that its function is to aid in maintaining apposition of the condyle–disc assembly on the articular eminence during functional movement. In view of the very small bulk of the upper head and the small amount of power it can generate, its mechanical role in assisting elevation of the mandible would seem to be questionable. A stabilizing, guiding role would seem much more in keeping with the size and position of this small, flat, muscle mass (Koritzer and Suarez, 1980). Recent electromyographic studies on humans tend to confirm this proposition (Juniper, 1981; Koole et al., 1984; Widmalm, Lilie and Ash, 1987).

The digastric muscle

This muscle is formed by two belly-like masses of muscle tissue joined by an intermediate tendon (Figure 31 and 32).

The anterior belly is smaller and arises from the digastric fossa of the mandible, lateral to the mental symphysis. Its fibres course posteriorly, laterally and almost horizontally, on the superficial surface of the mylohyoid muscle and deep to the inferior border of the mandible, to the intermediate tendon. The muscle is located in an elastic, investing sleeve of the deep cervical fascia which holds it laterally close to the deep surface of the lower border of the mandible. It is a belly muscle and its maximum cross-section is opposite the second molar region (Widmalm, 1988; Figures 31 and 32).

The fibres of the anterior belly are supplied with motor, proprioceptor and exteroceptor innervation by a branch of the mylohyoid branch of the mandibular division of the fifth cranial nerve.

The posterior belly arises from the mastoid notch of the temporal bone. The fibres course anteriorly, medially and slightly inferiorly to the intermediate tendon. This is a larger muscle mass than the anterior belly (Figures 31 and 32).

The level and orientation of this muscle, like that of most of the mandibular muscles, is often presented inaccurately by descriptive anatomists for illustrative purposes.

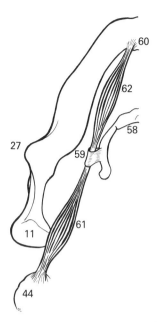

Figure 31 Digastric muscle in horizontal perspective. The possibility that these muscles can guide the mandible medio-laterally is evident in this perspective which shows the laterally-placed posterior attachments and the medially-placed anterior attachments: (11) condyle; (27) coronoid process; (44) mastoid process; (58) hyoid bone; (59) intermediate tendon of digastric fascial sling; (60) digastric fossa; (61) posterior belly of digastric muscle; (62) anterior belly of digastric muscle

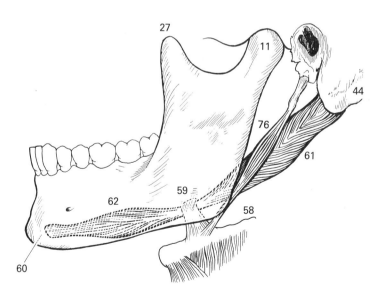

Figure 32 Diagrammatic representation of the digastric muscle in sagittal perspective. For illustrative purposes the hyoid bone is drawn more caudal than it is actually placed. Code as in Figure 31 apart from: (76) stylohyoid muscle

The fibres of the posterior belly are supplied with motor and proprioceptor innervation by the digastric branch of the seventh cranial nerve.

Mechanical features

The intermediate tendon of the digastric muscle is housed in a fascial tunnel formed from a loop of the cervical fascia which is attached to the hyoid bone. This fascial tunnel, in conjunction with a fixed hyoid bone, forms a pulley through which the intermediate tendon can slide to and fro. This enables the digastric muscle to exert inferior and posterior traction, as well as a lateral vector on the mental symphysis (Koritzer, 1983).

Contraction of the digastric muscle applies traction to the body of the mandible near the midline, in a posterior and inferior direction. If the anterior and posterior bellies are considered to act as one, then the mechanical advantage of the digastric muscle is the greatest of all for achievement of the depression of the mandible. Its activity as a depressor of the mandible in functional movements is well established (Kawamura, 1968; Vitti and Basamajian, 1977; DuBrul, 1980).

The posterior and intermediate attachments of the digastric muscle are situated more laterally than the anterior attachment which is close to the midline. The anterior belly of the digastric is a thin, weak belly of tissue, often fused with the mylohyoid muscle. It is speculated that the two digastrics, acting together, may provide centring guidance for the upward moving mandible as it approaches the intercusping position in normal function. The two muscles, as part of their controlling antagonistic action to the elevators, have the capability of directing the mandible both laterally and medially, like the reins guiding a horse (Koritzer, 1983; Figure 31).

Functional activity of the digastric muscles

The greatest activity of the digastric muscles occurs during depression of the jaw. The right and left muscles contract synchronously, and with increasing vigour, in moving the mandible from the postural position to wide opening (Carlsöö, 1956; Moffett, 1977; Munro, 1972; Vitti and Basmajian, 1977; Widmalm, Lillie and Ash, 1988). The activity gradually decreases during closure. Thus the digastrics are the main depressors of the mandible and provide antagonism for controlled closure.

With the mandible stabilized in occlusion, the digastrics are active in assisting elevation of the hyoid bone during swallowing. The level of activity during this function is variable (Hrycyshn and Basmajian, 1972; Widmalm, Lillie and Ash, 1988).

The digastrics are active during all horizontal positioning movements of the jaw in right and left lateral, protrusive and retrusive excursions (Møller, 1966; Munro, 1972; Vitti and Basmajian, 1977; Widmalm, Lillie and Ash, 1988). They appear to contribute as prime movers and antagonistic controllers in horizontal positioning.

Widmalm and co-workers (1988) noted considerable activity in the digastrics during simulated gnashing, but no activity during static clenching of the teeth.

Summary
1. The digastric muscles have the mechanical capability to depress the mandible, raise the hyoid bone and influence the horizontal component of jaw movements.

2. The electromyographic evidence shows that the greatest activity occurs in the digastrics when they are acting as prime movers during depression of the jaw and that they are bilaterally active during right and left lateral, as well as during protrusive and retrusive, excursions of the jaw.
3. The digastrics, like the lateral pterygoids, participate in all functional movements of the jaw.
4. The digastrics are very active during gnashing movements of the jaw.
5. The digastrics are inactive during static clenching of the jaw.

The suprahyoid muscles other than the digastric

The remaining suprahyoid muscles attached between the hyoid bone and the mandible are the genioglossus, the geniohyoid and the mylohoid. These muscles may contribute to the depression of the mandible at times, but are probably not very significant in this respect. These muscles are movers of the hyoid bone and elevators of the floor of the mouth and tongue when the mandible is stabilized (Kawamura, 1968). With the hyoid bone fixed they may contribute to a small extent to depression of the mandible.

The infrahyoid group of muscles

This muscle tissue aids stabilization and movement of the hyoid bone and, as such, has a peripheral role in the function of the masticatory system.

Chapter 4

The remaining muscles of the head and neck

These muscles are primarily movers of the head and neck and their chief task is to control and maintain postural position. They are involved in the function of the masticatory system to a small extent — in the incision and tearing of portions of food and assimilation of food and drink. Kawamura and Kimura (1971) demonstrated the influence of head position on the activity of the mandibular elevators, e.g. rotation of the head interrupts the chewing rhythm and enhances the activity of the contralateral temporal muscle. The origin of these effects is believed to be the spindles of the neck muscles.

Kawamura (1974) also draws attention to the close relationship between the neuromuscular regulation of the activity of the pharynx and that of the masticatory system. Dysfunction of the styloid complex may upset pharyngeal function and, indirectly, masticatory system function (Koritzer and Kenyon, 1983).

The head and neck muscles can be involved in the general hyperactivity of the postural muscles and may be indirectly involved with dysfunction in the masticatory system (Rocabado, 1984).

Kortizer and Kenyon (1983) point out that the styloid process and its muscle complex is closely related to the craniomandibular joint; being positioned less than 10 mm postero-medial to it. There is continuity of fascia between the two structures. Inflammation or dysfunction in one area might effect the other (Figures 23 and 56).

The styloid musculature running antero-medially to the styloid process passes very close to the lateral edge of the transverse process of the atlas vertebra.

Arising from the back of the transverse process of the atlas, further down and passing upwards to insert into the basilar occipital bone, posteriorly, is the longissimus capitis which tips the head backwards.

The levator scapulae muscle arises from the back of the cervical vertebrae. Its fibres run inferiorly to the scapula.

The posterior belly of the digastric muscle arises from the mastoid notch and runs antero-inferiorly and medially to reach the hyoid bone. Superficially overlain by the sternomastoid muscle and the ramus of the mandible, the digastric muscle is related on its postero-medial side to the superior oblique and rectus capitus lateralis muscles and the atlas transverse process.

Koritzer and Kenyon (1983) emphasize the possible relationship between pain, inflammation and dysfunction in the masticatory system and similar disturbance in the post-cervical muscle–skeletal complex through the digastric and styloid areas.

Widmalm, Lillie and Ash (1988) also emphasize the close relationship between the posterior digastric and the deep muscles of the neck, especially the obliquus capitis superior.

Chapter 5
Biomechanical aspects of muscle function

Muscle tissue is a response to the need for power application in a multicellular organism. The tissue is located where it most usefully serves this function.

The named units of the tissue vary greatly in overall form, internal structure and the shape of the attachment to bone or other tissue. Each muscle is adapted to provide the necessary range, direction, and force of contraction to carry out the functions of the body system to which it belongs.

The physical activity of muscle is limited to contraction and the development of tension.

The general shape of a muscle and its internal arrangement of fasciculi is the result of numerous conflicting demands which have been reconciled to achieve the greatest efficiency of form and function in the given circumstances.

Where much power is needed the bulk of the muscle tissue will be large. However, excessive bulk can interfere with other structures in the neighbourhood. When the line of traction on a part is straight and the range of functional activity is limited, the attachments of the muscle can be reduced to confined areas, e.g. the geniohyoid muscle. When a large range of directional options are required, the attachments tend to cover an extensive area, e.g. the temporal and lateral pterygoid muscles.

When a lot of power is required but attachment is limited, e.g. masseter to zygomatic arch, the number of fibres is increased, but these are attached to internal fibrous septa which, in turn, are attached to the bone over an area much less than that required by muscle tissue.

When increased directional options are required, the fascicular alignment may be varied in many ways within the muscle, e.g. they may run in different directions in the same plane to give a cruciate arrangement (as in the masseter or temporal muscles viewed in lateral perspective). Some fibres may run almost straight, others take spiral path around them, e.g. the sternomastoid and trapesious muscles. It is possible that multipennate structure may also give directional options, e.g. the temporal muscle.

Muscle fibres range from 10 to 60 mµ in diameter and from 3 to 300 mm in length. In general, accurate and refined movements are carried out by the small fasciculi of small muscle cells, e.g. eyeball and jaw movements. Some muscle fibres may run the whole length of the muscle. Others may extend through part of the overall length, gaining insertion into tendons. Muscle shape and contour invariably blend, in a streamlined fashion, with the surrounding structures so that a fully adequate level of efficiency is provided with maximum compactness of form.

Most of the musculoskeletal system operates on the third class lever system. This is less efficient than the first or second class systems but allows a vastly more compact form to be achieved.

The maximum force which the muscle can develop is directly proportional to the mass of contractile tissue present.

The range of contraction of which the muscle is capable is directly proportional to the length of the muscle fibres and their positional relationship to the fulcrum of movement of the bone. In general, the range of contraction will be from one-third to one-half of the resting length of the muscle fibres.

When the fascicle alignment is the same as the general line of pull of the muscle, maximum efficiency of force is developed and the range of movement achieved is the greatest possible.

Chapter 6

Collective action of the muscles which move the mandible

Sicher and Tandler (1928) rationalized the description of the muscle function involved in various mandibular movements and their ideas are confirmed by modern anatomical and physiological investigation (Kawamura, 1968). Functional mandibular movement is characterized by its three-dimensional nature and the involvement of almost the entire area of muscle which moves the mandible during its performance (Pröschel, 1987; Møller, 1966).

Most of the prime movers of the mandible are attached to the skull at their other ends. Starting with the attachment of the temporal muscle, the elevators, protruders and retruders can be stripped or dissected away from the skull sides and base as a continuous mass of muscle. At certain points, however, there are endosteal tendinous attachments, e.g. on the zygomatic arch, mandible and pterygoid plates (Koritzer, 1983).

When the muscles are removed in this manner, the confluence of many neighbouring, but individually named, parts of the mass of muscle tissue becomes evident. This raises questions about the functional significance of such confluences (Koritzer and Kenyon, 1983).

Muscle slips are attached to the craniomandibular joint capsule or disc in a radial fashion (Koritzer and Keynon, 1983) (Figures 29 and 52).

Awareness of the likely significance of these anatomical observations is increasing and will be substantiated when electromyographic and other techniques are developed fully.

These small muscular entities have neither caused nor excited much enthusiasm among career anatomists. Dentists with a special interest in them are entitled to speculate on their probable significance, provided that the present lack of objective evidence to back up claims for suggested functions is borne in mind.

It is important to emphasize the fact that the elevators, retractors and protruders of the mandible can be 'peeled away' as a continuous unit from the skull sides and base. It would appear more sensible to consider this unit of variously aligned muscle fasciculi as a source of power at the disposal of the central nervous system, to be called upon to position and move the mandible according to the functional needs at any given time, rather than to segregate parts of it and ascribe names to them. In this way the comprehensive integration and involvement of all of this muscle tissue in most functional movements of the jaws will be better understood and reinforced (Koritzer, 1983).

Some characteristics of functional mandibular movements

The common functional movements of the mandible are characterized by their:

1. Short range. Seldom more than half of the range of the movements on the border of the envelope of movement (Hildebrand, 1931; Fischer, 1935; Posselt, 1952; Lundeen and Gibbs, 1982; Pröschel, 1987).
2. High velocity. Acceleration is greater than that due to gravity (Goodson and Johansen, 1975).
3. Occurrence through the three planes of space simultaneously (Hildebrand, 1931; Posselt, 1952; Lundeen and Gibbs, 1982; Pröschel, 1987).
4. Great precision, especially close to tooth contact, to avoid traumatic incidents in the masticatory system (Hildebrand, 1931; Tryde, Frydenberg and Brill, 1962; Watt, 1981; Lundeen and Gibbs, 1982).
5. Close association with the functional activity of the pharynx, larynx and muscle and with the skeletal system of the neck (Kawamura, 1968)

Executing movements of this nature involves the simultaneous activity, of varying and constantly changing extent, of numerous muscles, and at any one moment the following will be in action:

1. The prime movers – determining the direction of movement and the power developed at that instant.
2. Synergists – assisting the direction, power generation and speed of the movement as determined by the prime movers.
3. Antagonists to the prime movers and their synergists which enable the movement to progress uniformly and to be steady and precise.

For example, in the habitual depression of the mandible there will be activity of:

1. The chief depressors – the digastrics and inferior parts of the lateral pterygoid.
2. The infrahyoid strap muscles attached to the hyoid bone – to fix the hyoid bone.
3. The remaining suprahyoid muscles as synergists to the digastrics.
4. The elevators and retruders as antagonists.

Similarly, in other functional movements virtually all of the muscles attached to the mandible will be involved as prime movers or synergists, or as antagonists to help control the movement. This is because most mandibular movement is through the three planes of space; it involves sliding movements in the joints and usually has to be quite precise (Lundeen and Gibbs, 1982). Movements of the mandible as distinct from the movements of other bones cannot, at any time, be reduced simply to the line of the pull exerted by one prime mover or another.

Owing to its complex nature, mandibular movement requires close central nervous system control at all levels. A relatively large proportion of the higher centres and brain stem is concerned with this task (Penfield and Rasmussen, 1950).

From the previous review of the individual elevators and protruders the overall direction of their pull, when they act bilaterally and synchronously, may be summarized as upwards and forwards viz (Figure 33):

1. Masseter and medial pterygoid is upwards and forwards.
2. Lateral pterygoid is forwards and temporal is mainly upwards.

Therefore, the major vectors created by the mandibular muscles will all tend to seat the condyles in the close-pack position in the craniomandibular joint, or, when

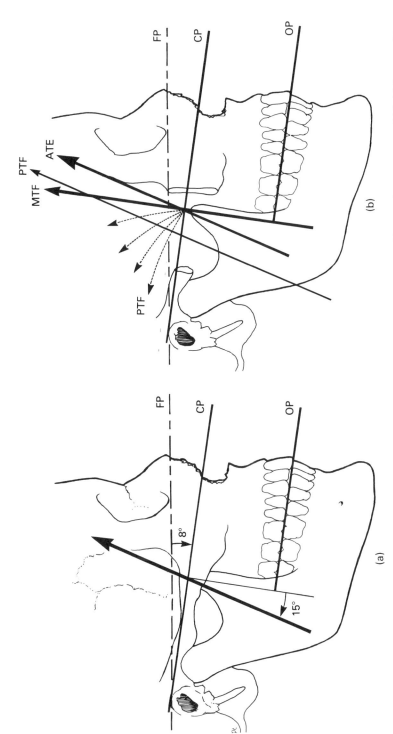

Figure 33 Diagram of the line of elevator traction (after Schumacher, 1961): (a) masseter and medial pterygoid – anterior component of 10–15 degrees with perpendicular to occlusal plane; (b) temporal (most anterior fasciculi) – anterior component of 10 degrees with perpendicular to occlusal plane (main bulk of retro-orbital temporal fibres perpendicular to occlusal plane); (FP) Frankfort plane; (PTF) posterior temporal fibres – pull up with anterior component when the mandible is close to maximum intercusping position; (ATF) anterior temporal fibres; (MTF) main bulk of retro-orbital temporal fibres; (CP) Camper's plane

the condyle-disc assembly is out of the glenoid fossa, hold the condyle-disc assembly in contact with the distal surface of the articular eminence.

Even in the postural position this tendency will be evident (DuBrul and Menekratis, 1981). The depressors of the mandible would appear to augment the result of the elevators or protruders in achieving the above effect. The suprahyoid group, attaching anteriorly to the body of the mandible, forms a force-couple (DuBrul, 1980) with the elevators which tends to rotate the condyles into contact with the distal surface of the articular eminence. This principle can be incorporated into clinical manipulation techniques for moving the mandible.

Summary of named areas of mandibular muscle involved in basic mandibular movments as prime movers

Elevation

Synchronous, bilateral activity of masseters, medial pterygoids and temporals with coordinated antagonism from the suprahyoid group. Coordinated activity of the superior part of the lateral pterygoid muscle.

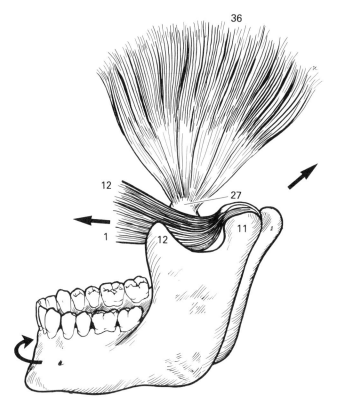

Figure 34 Diagrammatic representation of the prime movers for the lateral jaw – non-working side lower head of the lateral pterygoid, non-working side of the medial pterygoid (not shown) and working side rotation of the posterior temporal fibres. The elevators and depressors stabilize the jaw at the required vertical height for lateral movement: (1) inferior part of lateral pterygoid; (11) condyle; (12) superior part of lateral pterygoid; (27) coronoid process; (36) temporal muscle

Depression of the mandible

Synchronous, bilateral activity of the inferior parts of the lateral pterygoids and the digastrics and other suprahyoids, with coordinated antagonism from the 'elevators'. In general the movement is initiated by the lateral pterygoid and the digastrics come in later (Kawamura, 1968). Møller (1966), in a classical electromyographical study, found that the digastric was activated ahead of the lateral pterygoid muscle in functional depression, e.g. chewing.

Protrusion

Synchronous, bilateral activity of the inferior parts of the lateral pterygoids, helped by the masseter and medial pterygoid. The digastrics and posterior temporals supply an antagonistic effect.

Retrusion

Synchronous, bilateral activity of the posterior and middle temporal and digastrics and other suprahyoids. Synchronous, bilateral activity of the superior part of the lateral pterygoid controls disc retrusion in the craniomandibular joint. The inferior part of lateral pterygoid provides antagonism.

Lateral mandibular movement

This is achieved by coordinated synchronous movement of the working-side temporal muscles and the non-working-side pterygoids, i.e. the medial ptergygoid and inferior part of the lateral pterygoid. The latter swings the mandible across the midline in the horizontal plane while the working-side temporal 'muscle' aids lateral swing and stabilises the working-side condyle, helping it to act as a pivot for the lateral movement. Coordinated action of the elevators and depressors of both sides sets the vertical plane at which the lateral movement occurs level (Figure 34; see also pp. 100–104).

Chapter 7

Evolutionary and embryological features of the prime movers of the mandible and the craniomandibular joint

Embryologically, most of the muscles which move the mandible develop from the mesenchyme of the first branchial arch.

These muscles have evolved from the adductor mandibulae of primitive animals such as fish. Phylogenetically, the complex of tissues caudal to the tooth carrying and producing area in primitive animals gradually evolved (Moffett, 1980) to form:

1. An auditory–sensory mechanism based on vibration and resonance.
2. A joint mechanism for the lower jaw.
3. Muscular tissue to move the lower jaw.

In the development of the human embryo, at about the 25 mm stage:

1. Meckel's cartilage extends from the midline of the mandible to the ear region on either side.
2. Mesenchmal tissue, which develps into the joint disc, extends through the future joint to the lateral pterygoid muscle which is medial to the masseter muscle above the cartilage (Moffett, 1984).
3. The zygomatic process of the temporal bone is above this mesenchyme.

The Meckel's cartilage and related areas develop into the middle ear ossicles, sphenomandibular ligament and spine of the sphenoid. The mandible develops lateral to Meckel's cartilage. The zygomatic process of the temporal bone forms the upper surface of the craniomandibular joint. Between the latter and the mandibular condyle, the disc, joint spaces, capsule and synovia are evolved by mesenchymal differentiation and breakdown (Symons, 1952; Moffett, 1984).

Continuity between the superior part of the lateral pterygoid muscle, the craniomandibular joint disc and the malleus of the middle ear can be established embryologically and in early life (Rees, 1954; Pinto, 1962; Moffett, 1980).

Phylogenetically, of the prime movers of the mandible the temporal muscle is the first to develop and is present in fish. The masseter muscle appears in animals at the phase of the reptilian – mammalian interface, and the horizontal positioners of the pterygoid complex were the last of the prime movers to evolve. However, the fact that the lateral pterygoid muscle is sometimes innervated by the auriculo-temporal nerve is taken by some to indicate a less specialized and more general early origin than is usually attributed to it (Koritzer and Suarez, 1980).

The common embryological origin of the major movers of the mandible is attested by the frequent partial unions or very close association of the named muscle masses which are found at dissection.

1. The masseter and temporal muscles – at the posterior part of the deep head of masseter (zygomatico-mandibular muscle).
2. The masseter and the medial pterygoid on the medial surface of the mandible (MacConaill, 1975).
3. The temporal and the superior head of the lateral pterygoid (Schumacher, 1961; Koritzer and Suarez, 1980). The lateral fibres of the latter are closely intermingled (Christensen, 1969) with the deep fibres of the temporal muscle at the infratemporal crest.
4. The perimysium and muscle fibres of the temporal and the deep head of the masseter often attach to the capsule and disc of the craniomandibular joint (Rees, 1954; Widmalm, Lillie and Ash, 1987).
5. The superior part of the lateral pterygoid muscle and the inferior part of the lateral pterygoid anterior to the insertion (Gray, 1858; Sicher, 1960).
6. The lateral ptergyoid muscle, disc of the craniomandibular joint and malleolar ligament of the middle ear (Rees, 1954; Pinto, 1962; Moffett, 1980).

The joints of the skull and masticatory system

Chapter 8

Craniofacial and diarthrodial joints in general

A joint is a point of union of two or more bones. The functions of joints are to participate in:

1. The tasks for which bones are intended; by forming a firm but resilient union between bones, or forming a union between the bones in such a way that movement can occur between them.
2. Growth and development of the skeleton.
3. Adaptive remodelling of the fully grown skeleton to change in its environment.
4. The transmission and dissipation of energy within the skeleton.

Classification

Joints may be classified as:
1. Synarthrodial joints. In these joints the bone surfaces are joined by white collagen bundles forming:
   (a) sutures – end-to-end union of the bones;
   (b) syndesmoses – union of the bones by an interosseous membrane or ligament, as between the tibia and fibula;
   (c) gomphosis – union of a conical process of bone into a socket, as teeth in alveolar bone (Greek = bolting together).
2. Diarthrodial joints. Such junctions are discontinuous and allow functional movement between the bones. There is no direct continuity across the joint surfaces by fixed fibrous tissue. The joint surfaces are able to move on each other. Diarthrodial joints are sub-classified as:
   (a) simple, i.e. involving only two articulating surfaces in the union;
   (b) compound, i.e. involving more than one pair of articulating surfaces and the articulating territory of each pair of surfaces remains quite distinct, e.g. elbow joint.

Craniofacial joints (after Moffett, 1980)

There are 209 craniofacial joints and they are classified anatomically as:

1. Synarthrodial (119):
   (a) gomphoses – tooth or alveolar bone in primary and permanent dentition;
   (b) sutural – facial, between facial bones and facial and cranial bones, and cranial between the bones of the cranium;

    (c) auditory – ligamentous joints of ear;

    (d) cartilagenous – cranial synchondroses, condylar cartilages and nasal septum.

2. Diarthrodial (8):

    (a) craniomandibular;

    (b) atlanto – occipital;

    (c) auditory ossicles.

3. Special dental (82):

    (a) occlusal between the maxillary and mandibular occlusal surfaces in primary and permanent dentitions;

    (b) interproximal between the proximal surfaces of adjacent teeth in each arch of teeth, in primary and permanent dentition.

See Table 8.1.

At times it is useful clinically to relate the craniofacial joints to each other in a sequential manner from the occlusal 'plane' outwards. In the occlusal classification of craniofacial joints (Moffett, 1980) there are four orders. The first order covers the occlusal joints, the second order the periodontal joints, the third order the maxillofacial suture and craniomandibular joints and the fourth order the cranial synarthrodial joints.

Energy or forces passing through occlusal surfaces are transmitted sequentially through the craniofacial joints. Energy or force diminishes in magnitude through each order of the joints.

By viewing the joints in reverse order the joint growth sites, which help to determine arch and tooth positions in each jaw, can be identified (Moffett, 1980).

Dental joints (proximal tooth contacts and occlusal contacts) are the only joints of the system which do not have a cellular remodelling mechanism. The form of these joints can be changed only by wear of the surfaces, disease and dentistry.

The location and form of these joints are important because they influence the function, efficiency and comfort of the masticatory system (Moffett, 1980) via the proprioceptor system of the joints – in the periodontal tissues. The absence of a cellular remodelling mechanism is significant because the only natural way of adapting to a change in the environment is by wear, i.e. tissue loss at the joint, and there is no biological mechanism which will repair or adapt to the effects of disease because lost tissue cannot be regenerated in the absence of a cellular mechanism (Moffett, 1980).

General features of diarthrodial joints

The functions of joints are to connect bones and/or to enable movement to occur between them. Synarthrodial joints enable skeletal parts to be assembled together as a continuous unit in a less rigid and more resilient way. Diarthrodial joints are mainly concerned with providing conditions under which bones can move in relation to each other. Some features commonly associated with all diarthrodial joints are described below.

Articular surfaces

Where surfaces move in relation to each other the contacting parts will be subjected to compressive and shearing forces and should be resistant to them. The tissue on

Table 8.1 A summary to some of the biological features of the craniofacial joints (Moffett, 1980)

Microanatomical classification	Morphological classification	Physiological functions	Mechanical mechanism by which function is carried out	Possibility of adaptive remodelling
Dental	(i) Occlusal and (ii) interproximal	(i) Mastication (ii) Speech (iii) Facial expression (iv) Transmission of energy/force	(i) Development of shearing force (ii) Compression	Nil: (i) Acellular tissue (ii) Avascular tissue
Synarthroidal Fibrous	(i) Periodontal (ii) Sutural	(i) Tooth eruption (ii) Cranial and facial growth and development (iii) Mastication (iv) Transmission/dissipation of energy and force (v) Adaptive remodelling (vi) Kinesthetic sensation	(i) Developmpent of tensile forces (ii) Compression	Very high (i) Cellular tissue (ii) Vascular tissue
Cartilaginous	(i) Condylar (ii) Nasal (iii) Synchrondrotic (iv) Branchail period	Growth + development (therefore transitory)	Compression (tension on these joints may stimulate growth)	Limited (i) Cellular (ii) Avascular during growth period
Diarthrodial	(i) Craniomandibular (ii) Atlanto-occuipital (iii) Auditory	(i) Mandibular movement (ii) Head movement (iii) Neck movement (iv) Auditory conduction (v) Sensory information of bone movement	(i) Highly lubricated (sliding, shearing) (ii) Minimal friction (iii) compression	Limited (i) Cellular (ii) Avascular

the articulating surfaces needs to be rigid and strong (to maintain its shape), resilient and elastic, tough enough to resist shearing and abrasion and as frictionless as possible.

Most commonly, white collagen is the chief tissue on the articular surface and in many situations is made more rigid and elastic by a cartilagenous matrix. Friction is greatly reduced by an albumen-like lubricant, synovial fluid. The actual articulating surface cannot be innervated or vascularized because of the compression it must sustain. The nutrition and metabolism of the articulating surface is maintained by the synovial fluid. Articular surfaces never have simple geometric forms. The degree of curvature of the surfaces changes at any point taken across any profile of the surface. Hence the full range of movement of the bone around a particular 'axis' is not fixed and stationary in space (Williams and Warwick, 1980).

Range of movement

Simple or compound joints are structured to permit various movements. The range of movement which can occur in a joint will be determined by the functions which it subserves.

The range of movement will be related to the ligamentous attachments of the joints, the morphology of the bony parts, the muscle activity and other structural restraints.

When movement of a bone at a joint is limited to rotation about a single axis the joint is termed ginglymoid – a hinge joint – and is said to possess one degree of freedom.

When independent movement can occur around two axes at right angles to each other the joint is termed biaxial and is said to possess two degrees of freedom.

When a bone at a joint can move around three areas at right angles to each other it is said to possess three degrees of freedom.

A bone that is able to rotate around two or more axes can usually rotate in many intermediate positions too, i.e. such joints are usually multiaxial.

Diarthrodial joints may permit joint surfaces to slide over each other. This linear movement is termed 'translation' and is usually associated with joints which permit large angulation changes, such as the shoulder and jaw joints.

A joint with three degrees of freedom and a sliding capability will require more muscular control than a joint limited to simple ginglymoid movement. In the latter, the joint contours relationships and ligamentous attachments greatly confine the movement range, e.g. finger joints.

The limitation of the direction and range of movement to functionally habitual areas helps the development of skilled control of movement and is linked with the most advantageous distribution of the available muscle power. The more the morphological and ligamentous factors can 'specialize' the range of movement the less the requirement for bulky muscular dispositions around the joints.

The capsule of diarthrodial joints

The capsule takes the form of a cuff which encloses the articular surfaces of the bones involved in the joint. Each end of the cuff attaches in a continuous line just beyond the articular surfaces of the joint.

The chief element consists of white collagenous fibres gathered into bundles which run mainly parallel to each other and which interlace.

Where translatory movement occurs, the capsular cuff fibres will be long enough to give slack in the capsule when the joint is in its normal postural position. This allows translatory movement. Blood vessels and nerves penetrate the capsule. The chief functions of the capsule are:

1. To support the synovial membrane.
2. To enclose the synovial fluid.
3. To support and stabilize the articulating parts.
4. To restrict joint movement.
5. To support neuronal end-organs and the joint blood supply.

Ligaments

A ligament is a band of white collagenous tissue specially developed and arranged to connect and strengthen joints and support visceral organs.

Ligaments are made up from parallel bundles of collagen fibres which usually are interlaced by other bundles or fibres of collagen. The cellular element is small and consists of flattened fibroblasts. Most ligaments are white and shiny in appearance.

Ligaments are tough and flexible, but exhibit little or no capability for extension when tensile force is applied.

The main function of ligaments is to connect and strengthen the joints. In forming the capsule of diarthrodial joints they also provide a base for the synovial tissues and the kinesthetic sensing organs of the joint capsule.

Ligaments are often compared with string or rope which is used to bind objects together. String will not hold things together unless it is tightened by an external force but, when it is tightened, it will function to hold parts together. With regard to their binding and supporting function lax ligaments are inactive and taut, ligaments are active. All biological material, including collagen bundles, has some elasticity. Thus collagen bundles can be made tight and then taut, like instrument strings (McConail and Basmajian, 1969). Further tightening may then rupture the collagen bundles or their attachment to the bone. In the close-pack position of a diarthrodial joint all of the main ligaments are taut. The close-pack position is that position in which the maximum area of the articular surface is in contact and firmly held together, and in which maximum loading occurs on the joint surfaces. In other than the close-pack position some of the ligaments are tight but none are taut, and it may be possible to distract the articular surfaces slightly. In other than the close-pack position there will be a considerably reduced area of articular surface contact (see also p. 52–54 and Chapter 12).

In most positions which a diarthrodial joint takes up the ligaments do not act to bind and support the joint surfaces and, if the muscles associated with the articulation are completely relaxed, the joint surfaces can be separated. Ligaments do not prevent the dislocation of joint parts, they simply limit dislocating tendencies during functional movements (McConail and Basmajian, 1969).

Capsular ligaments
Often the capsule will be thickened considerably at certain parts by extra bundles of parallel fibres. Such localized thickenings are called 'capsular ligaments'.

Accessory ligaments
The 'capsular ligaments' are to be distinguished from 'accessory ligaments' which stand clear of the joint capsule. Accessory ligaments may be intracapsular or extracapsular.

Protection of ligaments

All ligaments are designed to prevent excessive or abnormal movements. Every ligament becomes taut at the usual limit of some particular movement. Ligaments are protected against excessive tension by appropriate muscle activity which is programmed into the pattern generator for the movement and which is reinforced by instantly effective nociceptive reflex inhibition.

The synovial membrane

This is a pink, smooth, shiny membrane which lines the capsule and any other non-articular, intracapsular surface in the joint. The membrane does not extend onto any of the articular surfaces. The membrane may be folded in on itself inside a slack capsule and the synovial surface area may be further increased by the presence of finger-like projections (synovial villi).

The folds and villi form flexible masses of tissue which fill potential space when the joint moves and enable movement without tearing the synovia.

The membrane consists of a surface layer of synovial cells (one to four cells deep) in an intercellular matrix. Cell shape varies from fibroblast to polyhedral. The cells are classified into two types:

1. Type A. These predominate. They have surface filopodia, plasma membrane invaginations and vesicles. Neighbouring cells are often separated by gaps. They have phagocytic powers.
2. Type B. The characteristics of type A cells are poorly developed. These cells have a rough endoplasmic reticulum.

Replacement of surface layers is probably by transformation of primitive sub-surface cells.

Cell types intermediate between type A and B are plentiful. The described differences between the surface layer cells may simply reflect quantitative variations in functional activity rather than distinct cell lineage (Williams and Warwick, 1980).

The sub-intimal layer of the membrane consists of a richly vascularized, loose, areolar tissue with many lymphatics. The presence of yellow elastic tissue prevents the formation of excessively redundant folds of the synovial membrane when the joint is in its resting position.

The synovial fluid

This is a clear or pale-yellow, viscous, glairy fluid carrying a mixed population of cells and amorphous particles. It looks like egg albumen, hence the name (McConail, 1950).

Its viscosity varies inversely with the shear stress applied to it, i.e. with low rates of shear it is quite viscous, with high rates of shear it is much less viscous.

Elasticity varies directly with shear stress. Its volume is always scanty, e.g. 0.5 ml for the knee joint.

It is a dialysate of blood plasma with some protein and added mucin (Wlliams and Warwick, 1980).

The cells in the fluid consist of monocytes, lymphocytes, macrophages, free synovial cells and some polymorphonuclear leucocytes.

The functions of the fluid are to provide a liquid environment for the articular surface of the joint, provide a nutritive source for the articular surfaces, to act as a lubricant to reduce friction on the articular surfaces and to remove debris and catabolic products from the joint cavity.

Circulation of synovial fluid
This is a controversial topic. There are many hypotheses, the four chief ones being:

1. Hydrodynamic lubrication. The articular surfaces are soaked in lubricant which forms a film preventing contact between the moving parts, like engine oil in a car engine.
2. Boundary lubrication. The synovial fluid bonds chemically with the articular surface to provide a nearly frictionless surface.
3. Weeping lubrication. The fluid fills the porous articular surface, e.g. cartilage. The application of pressure to the slightly compressible elastic surface squeezes lubricant fluid onto the surface to provide a frictionless layer.
4. Boosted lubrication. According to this theory, the articular surfaces are slightly deformed under compression. As a result of this, the synovial fluid collects in the depressions created. Further compression forces the small molecular components into the porous surface. The large molecular components that remain have improved lubrication properties.

Because of the wide variation in joint morphology, geometry and activity it is probable that different mechanisms operate under different conditions. The thickness of the fluid layer on each articular surface is thought to be about 10 µm, (Williams and Warwick, 1980).

Movement and stabilization in diarthrodial joints

Diarthrodial joints are pressure-bearing joints. The articular surfaces are in contact under varying pressure at all times. The surface contact is maintained by the activity of the muscles which pull across the joint (Sicher, 1960).
    The muscle tension maintaining contact is often aided by joint morphology and the ligamentous arrangement, e.g. close-pack position.
    Because of the requirement for movement most diarthrodial bony joint parts are only an approximate, loose fit in most positions taken up by the joint parts.

The close-pack position of diarthrodial joints
Only in one position do the articular surfaces fit together almost perfectly. This is called the 'close-pack' position of the joint.
    The 'close-pack' position has the following characteristics:

1. It usually occurs at one extreme of the most habitual movement of the joint.
2. The articulating surfaces are fully congruent.
3. The area of articular surface contact is maximal for that joint.
4. There is maximum loading on the articular surfaces.
5. The articular surfaces cannot be distracted; the articulating bones can be regarded as being locked together.
6. Enormous stresses can be generated.
7. The articular surfaces can be traumatized.
8. The fibrous capsule and ligaments may be maximally twisted and tense.

9. Force which tends to change bone relationships in this position is actively resisted by the musculature attached to the bones.
10. This musculature will act similarily in positions near the 'close-pack' position.

In all other joint positions the articular surfaces are not maximally congruent and the capsule is lax in some parts. These are 'loose-pack' positions in which the articular surfaces have a less stable relationship and can be distracted in many cases.

The approximate fit of articular surfaces in many diarthrodial joints allows rotatory and sliding movements and means that the surface parts in contact are small and constantly changing (this reduces friction and erosion during movement). It also creates spaces around the momentary articular contact which are filled with synovial fluid and may be important in synovial fluid circulation. Furthermore, as combined rotation and sliding can take place, the effective range of movement is greatly increased.

The limitation of movement in diarthrodial joints
The chief factors in limiting movement in these joints are:

1. The tension of antagonistic muscles – either the passive, elastic component of muscle tissue itself or the active contraction of the antagonist.
2. The tension in the joint ligaments and the accessory ligaments.
3. The approximation of soft parts (or teeth) or the compression of soft parts.

Summary

1. Diarthrodial joints are discontinuous articulations which allow movement of the bony parts.
2. Movement requires that articulating surfaces be hard, resilient, tough and frictionless.
3. The range of movement in the joint will be related to the habitual functional movement.
4. The smaller the range and type of movement the more specialized the morphology of the bony parts and the nature of the ligamentous arrangement.
5. The larger the range and type of movement the less specific the morphological and ligamentous arrangements of the joint and the more complex the muscular attachments and control.
6. All diarthrodial joints are encapsulated and provided with a synovial membrane and synovial fluid.
7. All diarthrodial joints have highly-developed kinesthetic sensing end-organs which provide a continuous input to the central nervous system pattern generator controlling the muscles which move the joint parts.

Chapter 9

The craniomandibular joint

Osteology

The bony parts of the craniomandibular joints are made up of the condylar process of the mandible, the glenoid structures on the skull base and the relationship between these bony parts and the dry skull (Figures 40 and 41).

The condylar process of the mandible

The posterior of the two upward projections from the body constitutes the condylar process. Posteriorly, on the ramus, this process is best positioned to act as a pivot for mandibular movement and as a fulcrum for a third class lever.

RIGHT

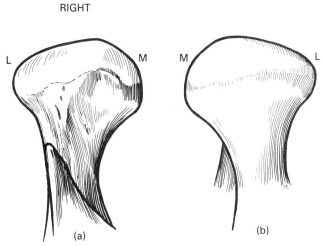

(a) (b)

Figure 35 Condylar process of mandible from anterior and posterior. Note the greater prominence of the medial pole and the marked concavity of the fovea on the anterior of the neck. The main articular facet faces antero-superiorly, extends lower medially than laterally and is much smaller than the posterior joint surface. The attachment of the joint capsule around the articular surface is indicated by the pencil line. A well-defined lip usually delineates the pterygoid fovea from the articular facet. The condyle is 15–20 mm long from pole to pole: (a) anterior view; (b) posterior view; (L) lateral; (M) medial

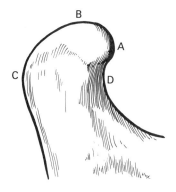

Figure 36 Lateral view of condylar process. This condyle is strongly convex along the top in this view. The main articular facet faces antero-superiorly and is often delineated (by a linear crest) from the posterior surface. The posterior surface included within the joint capsule is much greater than the antero-superior facet. The condyle has a maximum width of approximately 10 mm in this view. All surfaces curve away from the main articular facet: (A) inferior lip on main articular facet of condyle – the disc and capsule are attached at this line; (B) crest of condyle; (C) posterior line of attachment of capsule–disc assembly; (D) pterygoid fovea; (A–B) main articular facet, approximately 5 mm long; (B–C) posterior join surface of condyle, approximately 15 mm long

The condylar process is made up of a slim neck, flattened in the frontal plane, and a semi-cylindroid condyle (Figures 35 and 36). The condyle is about 20 mm long from pole to pole and 10 mm thick. The long axis is set latero-medially, approximately at right angles to the plane of the ramus (Figures 21, 29 and 37)

The posterior of the condyle is convexly rounded, both latero-medially and supero-inferiorly. The posterior surface faces postero-superiorly due to the anterior inclination of the whole condylar process. It is about 15 mm long, supero-inferiorly (Figure 36).

The rough antero-superior surface is slightly convex in the antero-posterior dimension and slightly convex or straight, latero-medially (Figures 35 and 36). It is usually about 5 mm long, supero-inferiorly. The antero-superior and posterior surface may form a crest-like junction, the highest point on the condyle. Often a slight sagittally-directed crest or groove separates the upper surface of the condyle into shorter lateral and longer medial parts.

There is great variability in the form of the condyle, even between the two sides in the same subject. At times the articular facet may slope sharply up from the poles to a high central 'peak' on the condyle (Meyer, 1865).

The condyle is extended medially and laterally into the medial and lateral poles. The lateral pole is rough, more pointed and extends less beyond the lateral surface of the ramus than the smoother and usually strongly rounded medial pole (Figures 35 and 37).

The latter extends well medial to the medial surface of the ramus. The long axis of the condyle is aligned so that, viewed in sagittal or horizontal perspective, the lateral pole is more anterior than the medial. The long axis of the condyles on the right and left sides, when extended medially, would meet posteriorly in the area of the anterior part of the foramen magnum and would make an anteriorly facing angle of about 150 degrees (Figure 37). The long axes of the condyles are

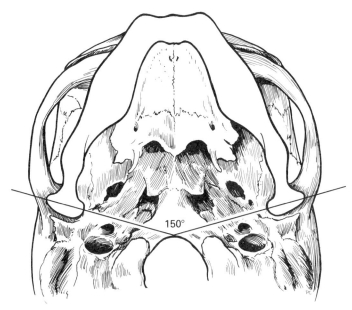

Figure 37 View of the condyles and fossae articulated in the dry skull. The long axes of the condyles and the glenoid fossae meet in the region of the foramen magnum, when projected medially, and make an angle of 150 degrees, open anteriorly

approximately parallel to the lines between the ipsilateral buccal and lingual tips of the tooth cusp (Kraus, Jordan and Abrams, 1969).

The antero-superior surface of the condyle is the articular surface. It slopes approximately to face the articular eminence, being helped in this by the anterior tilt of the whole condylar process on the ramus of the mandible. The articular surface stretches across the whole breadth of the condyle and down to the prominence of the medial pole (Figure 31). It is about 5 mm wide. It ends inferiorly, in a well-defined lip which overhangs the concave pterygoid fovea of the neck of the condylar process (Figure 35 and 36). All surfaces of the condyle curve away from the articular surface.

Summary

1. The condylar process of the mandible is designed:
   (i)   to support and position the articular surface of the mandible so that it can act efficiently, as a pivot for three-dimensional movement and as a fulcrum for forceful third class leverage;
   (ii) to provide a suitable attachment point for the muscle tissue which protracts the mandible.
2. To fulfil the above functions:
   (i)   the articular surface is positioned antero-superiorly and suitably shaped;
   (ii) all the other surfaces are inclined away from the articular surface so as to maximize freedom of movement in the joint, rigid support for articulation and convenience of muscle attachment (see also Chapter 11).

The glenoid structures on the skull base

The joint area of the skull base is housed in the squamous portion of the temporal bone just anterior to the tympanic part, lateral to the petrous part and posterior to the root of the zygomatic process. It consists of a vault-like hollow posteriorly, the glenoid fossa, and a slope rising out of the fossa anteriorly, the glenoid eminence.

The glenoid fossa
The glenoid fossa is concave latero-medially and antero-posteriorly.

It is much wider latero-medially than antero-posteriorly (Meyer, 1865). The long aixs of the oblong fossa is aligned postero-medially, like the condyle (Figures 37 and 38). In the deepest part of the fossa the bone is very thin and could not support the mandible (Figure 39) (DuBrul and Menekratis, 1981; Bell, 1982). The articular surface continues beyond the crest of the glenoid eminence onto the upward sloping infratemporal surface of the skull. The articular area here is called the 'preglenoid plane' (DuBrul, 1980; see p. 75). The lateral part of the fossa is wider than the medial part (Meyer, 1865; Figures 38 and 41).

The glenoid/articular eminence
The glenoid or articular eminence arising out of the fossa is strongly convex antero-posteriorly and slightly concave latero-medially, i.e. saddle-shaped, to correspond to the general shape of the articular part of the condyle (Figure 38).

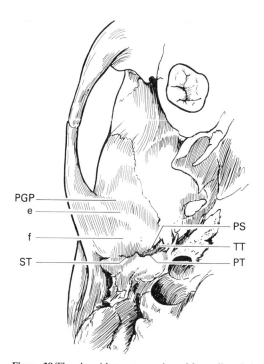

Figure 38 The glenoid structures viewed from directly below. The glenoid fossa (f) and eminence (e) are the main features. Their relationship to the zygomatic arch, anteriorly, and to the tympanic area, posteriorly, is obvious. Note the sutures – squamo-tympanic (ST), petro-squamosal (PS) and petro-tympanic (PT): (PGP) preglenoid plane; (TT) tegmen typani

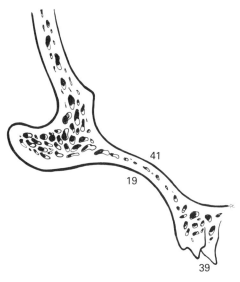

Figure 39 Frontal section through the glenoid fossa: (19) roof of fossa; (39) entoglenoid process/medial glenoid plane; (41) medial cranial fossa

The glenoid eminence runs down to a crest after which the surface turns upwards again gradually, the preglenoid area of the infratemporal surface of the skull base (DuBrul and Menekratis, 1981; Figures 38 and 41).

The medial and lateral borders of the eminance are sometimes accentuated by fine bony ridges. Anteriorly, the boundary of the joint is not usually marked on the dried skull (Sicher, 1960; Figure 53, p. 89).

The division between the squamous and tympanic parts of the temporal bone is evident in the glenoid fossa, laterally, as the squamo-tympanic suture (tympano-squamosal suture). Medially, part of the tegmen tympani bone protrudes between the squamous and other parts of the temporal bone dividing the fissure which exists between them into anterior (petro-squamosal) and posterior (petro-tympanic) parts, the latter sometimes called the Glaserian suture (Sicher, 1951; Figure 38).

The posterior of the fossa is raised to form a crest of bone, the posterior articular ridge (Sicher, 1951). Laterally, this ridge develops greater prominence to form a thickened cone in front of the external auditory meatus; the postglenoid process (Sicher, 1951).

Laterally, the border of the fossa is outlined by a small crest of bone running from the postglenoid process to the tubercle of the zygomatic arch, anteriorly (Sicher, 1951).

The oblong glenoid fossa is narrower medially than laterally. The medial side of the fossa is bounded by a downward extension of the squamous-temporal bone, the medial glenoid plane, to the medial of which is the spine of the sphenoid bone. The medial glenoid plane may extend down to form the entoglenoid process (Figure 39; Sicher, 1951).

The articular surface of the temporal bone is made up of the following parts:

1. The posterior slope and crest of the glenoid (articular) eminence.

2. The medial glenoid plane and entoglenoid process.
3. The preglenoid plane on the infratemporal surface of the squamous temporal bone.

Note that the roof of the glenoid fossa and the tubercle of the zygomatic arch are not articulating areas of the craniomandibular joint. The articular surface does not extend out to the lateral aspect of the root of the zygomatic process, as seen when viewed from the lateral aspect. The articular surface is several millimetres medial to the lateral surface and the articular surface is not as steeply inclined as is this lateral edge (Bell, 1982; Figures 37, 38 and 40). It is the latero-inferior border of the root of the zygomatic process and tubercle of the zygomatic arch which are seen on radiographs of the craniomandibular joint in oblique, lateral, transcranial views.

Summary
The glenoid structures of the skull base are designed to form the fixed articulating parts of the craniomandibular joint.

They are situated on the inferior surface of the squamous part of the temporal bone, surrounded by the tympanic and petrous parts of the same bone.

They take the form of an oblong, tapering, vault-like depression, the glenoid fossa, which roughly corresponds to the shape of the mandibular condyle, a well-defined eminence, rising anteriorly out of the fossa, which provides a downward guide for the anteriorly moving condyle and, on the medial side, a 'guiding ridge' of bone, the medial glenoid plane.

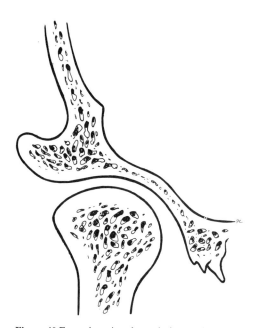

Figure 40 Frontal section through the craniomandibular joint at the level of the entoglenoid process. This illustrates the relationship between the condyle and fossa with the mandible in its maximum intercusping position (Krogh-Poulsen and Molhave, 1957; DuBrul, 1980)

The relationship of the bony parts of the craniomandibular joints in the dried skull

If the mandible is hand articulated to the dried skull in a fully dentate specimen, so that the teeth are in the maximum intercusping position, the relationship of the bony parts will show that:

1. A considerable discrepancy exists between the shapes of the articulating surfaces. An irregular space exists between the condyle and the glenoid structures which is usually greatest distally and medially. The condyle is only a very approximate fit in the fossa.
2. The condyle tilts forwards towards the posterior slope of the glenoid eminence (Figure 41).
3. The smallest space is that between the articular surfaces, laterally (Figures 40 and 41).
4. The lateral condylar pole juts out beyond the bone of the tympanic part of the temporal bone, posteriorly. In the fleshed skull this part of the condyle is masked by the cartilagenous part of the external auditory meatus and the tragus of the ear.
5. It is worth noting the area of the eminence down to the crest that is left uncovered by the condyle and the relationship of the medial pole to the medial glenoid plane.

In the fleshed skull the space between the bony components of the joint is filled with the soft tissue of the joint.

Summary
Relating the unfleshed bony parts of the joint with the mandible in maximum intercuspation demonstrates:

1. Considerable discrepancy in the amount of space between various parts; the condyle is only a very approximate fit in the fossa.

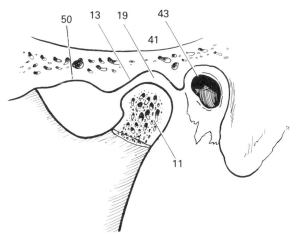

Figure 41 Sagittal view of bony relationships in the craniomandibular joint in the close-pack position (after DuBrul, 1980). The smallest space is between the antero-superior condylar facet and the postero-superior facet of the eminence in the lateral side of the joint. The lateral surface of the eminence and condyle have been cut away: (11) condyle; (13) articular eminence; (19) glenoid fossa; (41) middle cranial fossa; (43) external auditory meatus (50) preglenoid plane

2. The articular surfaces approximate to each other most closely in the lateral part of the joint.
3. The greatest amount of space between the bony parts is postero-superiorly and medially.
4. The lateral pole and postero-lateral aspect of the condyle project beyond the bony confines of the fossa.
5. The outline of the enclosed joint area on the skull base is marked nearly all the way around by a small crest of bone.

The soft tissue components of the individual craniomandibular joints

The soft tissue components of the craniomandibular joints are comprised of:

1. The articular covering of the joint surfaces.
2. The joint disc and attached muscle tissue.
3. The joint capsule.
4. The ligaments of the craniomandibular joint.
5. The synovial membranes.
6. Blood vessels and nerves.

The covering of the articulating surfaces of the joints

The surfaces of the joints which are subjected directly to load-bearing pressure are provided with a very smooth, specialized, tough, avascular layer of fibrous tissue which is highly resistant to rubbing and shearing stress and tightly bound to the bone surface. It is thick on the slope and crest of the glenoid eminence, the medial glenoid and preglenoid planes, and on the articular surface of the condyle, including that part down to the medial pole (Sicher, 1951; Neff and Suarez, 1983).

It is not present on the vaults of the glenoid fossae. There only a thin periosteum is present.

Microanatomy

At a microscopic level the articular soft tissue is seen to consist of a framework of collagen fibres interspersed with ground substance, some yellow elastic fibres, fibroblasts and cartilage-type cells. The fibres are arranged in the form of arches which are aligned antero-posteriorly and anchored in bone at both ends. Fine reconstruction and strengthening of individual collagen fibres occurs continuously (Steinhardt, 1958). The fibroblasts are elongated, spindle-shaped cells. Blade-like extensions of the fibroblast membrane extend between adjacent fasciculi of collagen bundles. The fibroblasts have an irregular stellate shape in cross-section. This ultra-microscopic picture clearly implies a vital relationship between these cells and the collagen bundles. The collagen, elastic and cellular elements are embedded in an amorphous, sulphur-containing mucopolysaccharide which cements the elements together (Ogus and Toller, 1986).

A thin, semi-calcified, chondroid zone exists between the collagen and the cortical bone. This area has cartilage cells which are most often evident on the crest of the articular eminence (Mahan, 1980).

Summary
The joint surfaces subjected to pressure are covered with a surface layer of densely inter-woven collagen bundles aligned antero-posteriorly, with fibroblasts and some elastic tissue, and a sub-surface, chondroid layer over the cortical bone.

The tough, avascular cover seems to be an adaptation to the intermittent shearing type of force to which the joint surface is subjected. The specialized surface is thickest where the stress is greatest and most often applied (Bell, 1982; Sicher, 1951).

The articular disc of the craniomandibular joint

Descriptive macroanatomy
The space between the bony and articular surfaces of the joint is occupied by a continuous structure named the articular disc (meniscus).

The disc is closely adapted to the shape of this space as it presents with the mandible in the maximum intercusping position, i.e. it fits snugly over the condyle and is closely adapted to the articular surfaces of the glenoid structure (Meyer, 1865; Neff and Suarez, 1983).

Viewed from above, the disc is oval or rectangular in shape with triangular flaps extending down, medially and laterally, to attach to the poles of the condyle (Kraus, Jordan and Abrams, 1969; Mahan, 1980; Figures 42 and 43). In the superior view, the disc takes on a pear-shaped outline the anterior fibrous extension of the disc to the lateral pterygoid muscle is taken into account (Rees, 1954) (Figures 51 and 52; pp. 86 and 87).

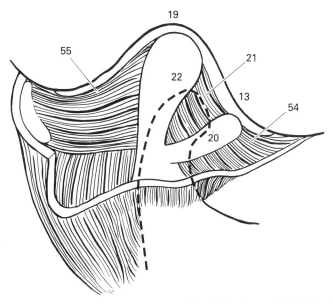

Figure 42 Disc of the craniomandibular joint (after Rees, 1954): (13) articular eminence; (19) glenoid fossa; (20) anterior band of disc; (21) intermediate zone of disc; (22) posterior band of disc; (54) anterior extension of disc; (55) bilaminar zone of disc

Viewed sagittally, it has the appearance of a schoolboy's skull cap (Rees, 1954; Figure 42).

Since it fits snugly over the condyle its inferior surface is oval medio-laterally and concave antero-posteriorly and medio-laterally.

The upper surface fits against the glenoid structures and hence, posteriorly, is convex in all directions to conform to the concave vault of the fossa. Anteriorly, the upper surface is convex latero-medially, but slightly concave antero-posteriorly, to conform to the glenoid eminence (Rees, 1954).

The thickness of the disc varies considerably in different parts antero-posteriorly. Rees (1954) describes three parts of varying thickness in the articular area (Figure 42):

1. The posterior band. This is the thickest part of the disc and the widest antero-posteriorly. Its upper surface is convex in all directions. Its inferior surface is concave in all directions. It fits into the vault of the fossa with the mandible in the position of maximum intercuspation. The superior crest of the condyle rests on the anterior part of the inferior surface of the posterior band.
2. The anterior band. This is a lesser thickening of the disc at the anterior end of the articular area. It is comparatively narrow anterio-posteriorly. Latero-medially, it is convex, superiorly, and concave, inferiorly. The condyle rotates on to this area in the latter phases of mandibular depression or protrusion (Rees, 1954; Krogh-Poulson and Molhave, 1957).
3. The intermediate zone between the anterior and posterior band. The disc is quite thin in this area. This is also a narrow zone of the disc antero-posteriorly.

The disc varies in thickness from its lateral to medial side. Thus in frontal section (Figures 40 and 43), it is seen to be relatively thin laterally and thick medially, corresponding with the usual dry skull relationship of the bony parts with the teeth intercusping maximally (Krogh-Poulsen and Molhave, 1957; Neff and Suarez, 1983).

Moffett (1984) has emphasized the self-seating form of the mature disc. He points out that the various thickenings of the disc surrounding the intermediate zone act mechanically as self-centring wedges. These automatically tend to keep the disc correctly positioned on the articular facet of the condyle when the healthy joint is loaded in function.

The posterior band is the thickest and most significant of these wedges and tends to prevent the condyle from being displaced dorsally off the disc or vice versa. The anterior band is a lesser thickening which also aids seating of the articular facet on the intermediate zone when the joint is under load at or close to its close-pack position (Figures 43 and 45).

The medial and lateral sides of the disc, in frontal section, show a self-centring, wedging arrangement too. The thickening on the medial side is marked due to the much bigger space between the bones on this side (p. 75; Figures 40 and 41). The intermediate zone and the wedging effect is thinnest in the lateral aspect of the joint. Thus it would be easier mechanically to displace the disc off the condyle from the lateral to the medial side (Figures 43 and 44).

As well as these articular areas of the disc, Rees (1954) describes two traction areas:

1. The anterior extension – a fibrous anterior extension of the anterior band to which the lateral pterygoid and other muscles attach, and which attaches to the anterior extremity of the preglenoid plane and to the anterior lip of the condylar

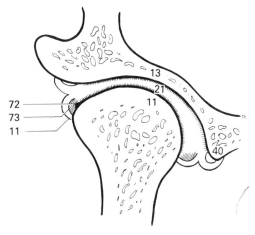

Figure 43 Frontal section through craniomandibular joint. The disc is thicker on its medial side (Krogh-Poulsen and Molhave, 1957). It presents self-centring thickenings laterally and medially (Moffett, 1984): (11) condyle capsule; (21) intermediate zone of disc; (40) medial glenoid plane; (72) collateral ligaments; (73) discal ligaments

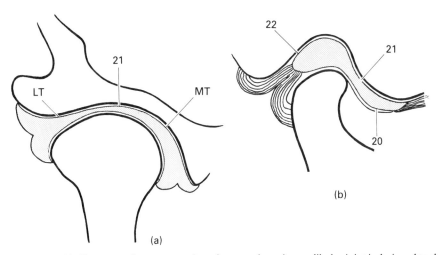

Figure 44 This diagrammatic representation of a normal craniomandibular joint is designed to show the thicker, more substantial dimensions of the disc around the intermediate zone laterally and medially (see frontal section a) and anteriorly and posteriorly (see sagittal section b). With the condyle pressed onto the intermediate zone, the thickened periphery has a wedge-like action which resists displacement of the disc off the condyle in the normal joint (Moffett, 1984). The lateral thickening is less than the medial and the anterior less than the posterior. The greatest wedging resistance to displacement of the disc off the condyle, in the anterior direction, will be from the postero-medial part of the disc: (LT) lateral thickening; (MT) medial thickening

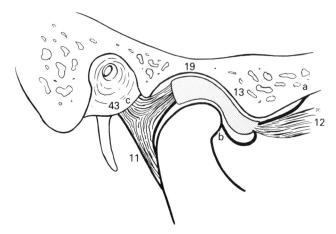

Figure 45 Attachments of the disc in sagittal perspective (after Rees, 1954): (11) posterior of capsule and condyle (also attachment of bilaminar tissue to condyle); (12) superior part of lateral pterygoid; (13) articular eminence; (19) glenoid fossa; (43) external auditory meatus; (a) attachment of anterior band to preglenoid plane; (b) attachment of anterior band to inferior lip of main articular surface of condyle; (c) attachment of bilaminar tissues to tympanic plate of temporal bone

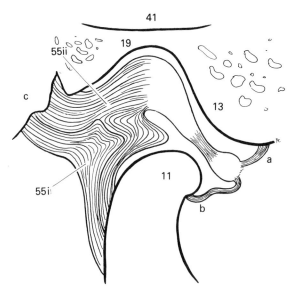

Figure 46 Sagittal view of craniomandibular joint with mandible depressed (after Rees, 1954): (11) condyle; (13) articular eminence; (19) glenoid fossa; (41) middle cranial fossa; (a) attachment of capsule–disc assembly to preglenoid plane; (b) attachment of anterior band of capsule–disc assembly to condyle; (c) attachment of bilaminar tissues/capsule to temporal bone; (55i) mainly white collagen tissue; (55ii) elastic tissue of bilaminar zone

process of the mandible. Traction via this extension exerts a pull in an anterior and medial direction on the disc (Figures 28 and 29).

2. The bilaminar zone. As the name implies there are two parts to this zone (Figures 45 and 46). The upper stratum is attached to the posterior wall of the glenoid fossa, up to and including the squamo-tympanic suture. This layer is composed mainly of elastic tissues (Figure 46). The lower stratum is composed chiefly of white collagen and attached to the inferior border of the posterior joint surface of the condyle (Figures 45 and 46).

These bilaminar tissues fold up against the back of the condyle, between it and the capsule at the posterior of the joint, when the condyle–disc assembly is seated in the fossa (Figure 45). As the condyle–disc assembly is protruded, the bilaminar tissues are pulled anteriorly with the disc. To accomplish this, the upper stratum, attached to the posterior wall of the fossa, must be extensible. Stretching the elastic tissue in the upper stratum develops a retrusive force to pull the disc back when the anterior traction ceases.

The fibres of the superior stratum consist of elastic tissue and are thick and strong (Rees, 1954). They are arranged in the form of a fenestrated membrane (Griffin and Sharpe, 1960).

The posterior mandibular attachment of the lower stratum is composed chiefly of white collagen fibres. This part of the bilaminar zone moves anteriorly with the condyle and does not need to be extensible. When the mandible rotates inferiorly, the mandibular attachment of the inferior stratum is rotated upwards towards its attachment to the temporal bone (Figure 48). Hence there is no need for elastic tissue at this site (Mahan, 1980). To create the necessary flexibilty between the superior stratum of the bilaminar zone and that of the inferior stratum and posterior part of the capsule of the joint, a layer of loose areolar connective tissue is interposed between them (Griffin and Sharpe, 1960). The attachment of the inferior stratum is low down on the posterior of the mandibular condyle just above that of the capsule of the joint at this point (Rees, 1954).

Summary

1. The disc occupies the space between the bony parts and articular surfaces of the joint.
2. Its general outline conforms to that space between the bones of the joint when the condyle is in the position dictated by the maximum intercusping position of the mandible.
3. The thin intermediate band of the disc is situated in contact with the articular surface of the condyle.
4. The thickest part, the posterior band, lies in the vault of the fossa over the postero-superior part of the condyle.
5. The thickened anterior band, with the mandible in its maximum intercusping position is situated down the slope of the eminence.
6. The anterior and posterior bands, together with the lateral and medial thickenings of the disc, form a self-centring mechanical wedging arrangement to maintain the disc on the condyle.
7. The anterior extension of the disc reaches to the crest of the eminence and is attached to the anterior of the preglenoid plane and the anterior lip of the articular surface of the condyle.

8. The tissues posterior to the condyle and the posterior band of the disc, the bilaminar zone, fold upon themselves and unfold as the condyle and disc are moved forward. The upper stratum is elastic and generates a retrusive force on the disc.

Microanatomical structure of the disc

There are three main areas to be considered under this heading: the articular areas; the vascularized parts of the disc; and the bilaminar zone (discussed above).

The articular areas   The anterior and posterior bands and intermediate zone are composed of densely-plaited, white, fibrous tissue (Rees, 1954). The collagenous bundles are orientated parallel to each other, antero-posteriorly, on the upper and lower surfaces and throughout the intermediate zone (Mahan, 1980) and here they are compressed more densely into a thinner space. The intermediate zone is completely avascular. In the anterior and posterior bands the collagen bundles are orientated in three dimensions and are bedded in a cartilaginous matrix (Griffin and Sharpe, 1960).

The cellular elements in the articular areas are mainly fibroblasts with occasional, irregular groups of round cells resembling cartilage cells (Rees, 1954; Moffett, 1984). The chrondrocytes are more numerous in older specimens and in the intermediate zone and posterior band (Griffin and Sharpe, 1960; Moffett, 1984).

The articular tissue is essentially dense, fibrous tissue, stiffened slightly by a cartilaginous-like matrix (Rees, 1954; Griffin and Sharpe, 1960, Moffett, 1984). This tissue is ideal for resisting the shifting, shearing type of forces to which it is subjected. Its toughness in this respect does not preclude adequate flexibility to enable it to adapt to differing degrees of tension and pressure. The lower content of cartilage-like matrix is an important factor in providing adequate flexibility and elasticity.

No blood vessels of any kind extend into the articular part of the healthy disc after the age of two-and-a-half years (Moffett, 1984). The cells in these avascular areas are maintained by the (passively) circulated synovial fluid.

There is no sensory innervation of any kind in the articular area of the disc (Moffett, 1984). The articular area can only wear out with use. It has no adaptive capacity because of its avascularity and acellularity (Moffett, 1984).

Summary

Histological examination shows the articular parts of the disc:

1. To consist of densely plaited, white fibrous tissue which is regularily structured and very tightly compacted in the intermediate zone and three-dimensionally structured in the anterior and posterior bands, ground substance which shows a chrondroid nature in the posterior band, fibroblast cells and occasional, irregularily-distributed cartilage cells.
2. To be completely avascular and non-innervated.
3. To have a structure which enables it to be tough and stiff while having considerable flexibility and elasticity.

The attachments of the disc

The articular disc is attached to:

1. The condyle.
2. The preglenoid plane.

3. The capsule of the craniomandibular joint.
4. The posterior wall of the glenoid fossa.
5. To muscles.
6. To the middle ear.

Attachment of the disc to the condyle   The disc is attached to the condyle anteriorly, posteriorly, medially and laterally.

Anteriorly, the disc is attached across the breadth of the condyle just below the ventral extremity of the articular area (Rees, 1954; Figure 47). This is a loose attachment to enable condylar movement on the inferior surface of the disc, i.e. the anterior extension projects far anterior to the articular area of the condyle; the attachment of the disc to below the lower lip of the articular area curls back 'on itself' (Figures 45 and 47). It consists of white collagen.

Posteriorly, the inferior stratum of the bilaminar zone is attached across the breadth of the inferior margin of the posterior slope of the condyle, below (Figures 35, 45 and 47), and the superior stratum is attached to the posterior band of the disc, above. The condylar attachment of the inferior stratum approaches the disc when the condyle is rotated on the inferior surface of the disc (Figures 46 and 48). The inferior stratum acts as a stopping mechanism to restrict forward rotation of the disc on the condyle.

Medially and laterally the disc is attached tightly to the poles of the condyle by triangular flaps of collagenous tissue 'discal ligaments', (Bell, 1982). The base of each flap is on the disc and the apex is attached just inferior to the pole of the condyle (Krogh-Poulsen and Molhave, 1957; Kraus, Jordan and Abrams, 1969; Figure 49). These attachments are such that the condyle can rotate on the inferior

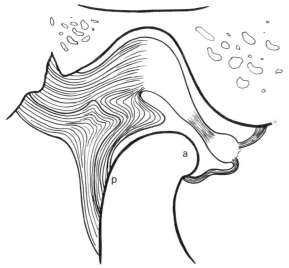

Figure 47 Attachments of disc to condyle: (a) anterior to or just inferior to the anterior lip of the articular facet; (p) posteriorly, just inferior to the posterior joint space. The posterior attachment rotates upwards towards the disc when the condyle rotates. The anterior attachment rotates down and away from the disc. Both of these attachments are slack and support synovial membrane on their inner surfaces

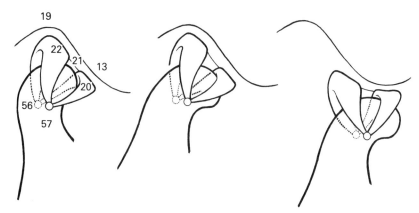

Figure 48 Attachment of disc to lateral and medial surfaces of the condyle, just inferior to the poles. This attachment is by strong collagen bundles gathered into triangular flaps (Kraus, Jordan and Abrams, 1969). The attachment binds the disc firmly to the condyle while at the same time permitting the condyle to rotate anteriorly on the inferior surface of the disc (after Rees, 1954): (13) articular eminence; (19) glenoid fossa; (20) anterior band of disc; (21) intermediate zone; (22) posterior band of disc; (56) attachment of disc at medial pole of condyle; (57) attachment of disc at lateral pole of condyle

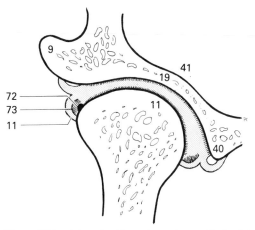

Figure 49 Frontal section through craniomandibular joint at poles of condyle to show attachment of disc to condyle. The drawing of the condyle is diagrammatic: (9) zygomatic arch; (11) condyle/capsule; (19) glenoid fossa; (40) medial glenoid plane; (41) middle cranial fossa; (72) collateral ligaments; (73) discal ligaments/triangular flap

surface of the disc in one plane only (Rees, 1954; Figures 47 and 48), displacement between condyle and disc is resisted and the disc is made to follow the condyle movements passively (Sicher, 1951).

The collagenous flaps are innervated and vascularized. Furthermore, the condyle can rotate very slightly around a vertical axis on the inferior surface of the disc (Kraus, Jordan and Abrams, 1969). However, the joint is designed to accommodate pivoting movement in the upper compartment between the glenoid articular surface and the superior surface of the disc.

The disc is tightly attached to the medial and lateral poles in the young healthy joint. No direct lateral movement of the condyle on the inferior surface of the disc is possible in the healthy joint (Mahan, 1980). The attachments are innervated and vascularized. Hence excessive stress on these attachments may cause pain and/or inflammation.

Summary

1. The disc is very firmly attached by white collagenous triangular flaps to just below the medial and lateral poles of the condyle. The arrangement enables the condyle to rotate freely, in the sagittal plane, on the inferior surface of the stationary disc. The disc may 'rotate backwards' relative to the condyle during condylar protrusion (Figure 47). Direct lateral movement does not occur in the lower joint compartment. The disc can be displaced from the condyle only by rupturing the attachment of these 'ligaments'.
2. The disc is attached to the condylar process posteriorly, below the distal slope of the condyle. This attachment moves with the condyle. It prevents anterior rotation or displacement of the disc on the condyle.

Attachment of the disc to the preglenoid plane   This attachment is to the bone at the anterior end of the articular surface of the preglenoid plane on the intratemporal surface of the skull (Figures 38 and 50).

   It is constituted by the fibrous or areolar tissue of the superior layers of the anterior extension of the disc (Rees, 1954). The anterior extension narrows anteriorly so that the latero-medial breadth of this attachment to the bone is narrower than the maximum width of the disc. This attachment is weak (Mahan, 1980).

Summary
The attachment of the disc to the preglenoid plane consists of fibrous areolar tissue unsupported by joint capsule. It is superior to the insertion of the lateral pterygoid muscle. It is a weak attachment.

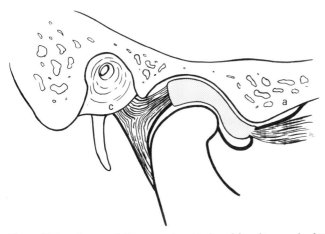

Figure 50 Attachment of disc to preglenoid plane (a) and tympanic plate (c)

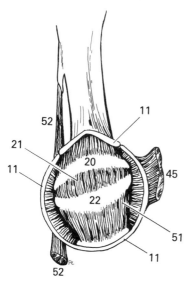

Figure 51 Upper compartment viewed from above (after Rees, 1954): (11) joint capsule; (20) anterior band of disc; (21) intermediate zone of disc; (22) posterior band of disc; (45) temporomandibular ligament; (51) menisco-capsule sulcus in upper joint compartment (upper compartment 'gutter'); (52) sphenomandibular ligament

Attachment of the disc to the capsule of the joint   Medially and laterally the disc is continuous with the capsule of the joint to form sealed and separate upper and lower joint cavities (Rees, 1954). While the disc is closely adapted to the condylar surface and tightly bound to the poles at the condyle, the medial and lateral walls of the capsule are lax and attached below the poles with the apices of the disc flaps. Medially and laterally the capsule–disc junction forms a deep sulcus (the upper compartment 'gutter') in the upper joint compartment which is evident in frontal section or in superior view of the upper compartment (Rees, 1954; Figure 51).

Posteriorly, the attachment of the disc to the capsule, medially and laterally, is continued into the superior stratum of the bilaminar zone.

Anteriorly, the capsule blends, medially and laterally, with the anterior extension of the disc. There is no well-defined capsule on the anterior of the joint (Rees, 1954; Mahan, 1980; Figure 50).

Summary

1. The disc–capsule attachment defines two joint cavities.
2. A deep medial and lateral sulcus in the upper joint cavity is developed by the junction.
3. The capsule of the joint is quite lax outside the disc attachment and does not restrain the disc movements.
4. The joint capsule is poorly defined, anteriorly.
5. The anterior attachment doubles back on itself to insert into the inferior lip of the articular surface. This enables 'posterior rotation' of the disc on the condyle (anterior rotation of the condyle on the disc) to be freely accomplished.
6. Excessive strain on the attachments may cause pain or inflammation.

Attachment of the disc to the posterior wall of the glenoid fossa   This attachment is through the superior stratum of the bilaminar zone. It consists mainly of thick fibres of elastic (Rees, 1954) arranged in a fenestrated structure (Griffin and Sharpe, 1960). These are strongly interwoven with the posterior band of the disc at one end and attached to the tympanic plate of the posterior wall of the fossa and squamo-tympanic suture at the other end. It is a bulky tissue which is folded upon itself in the posterior of the joint (Figures 46 and 50).

As this strong elastic band of tissue unfolds it enables the disc to be moved anteriorly and, when stretched, exerts tension on the disc. This helps to adjust its shape to the articular surfaces of the bone and provides a retrusive force to help reseat the disc in the fossa as anterior traction on the disc is reduced.

Summary
The superior stratum of the bilaminar zone is attached between the posterior band and the tympanic plate. It is primarily elastic. It is folded upon itself. The latter two features enable it to unfold anteriorly and be stretched as the disc is protruded. Its elastic recoil develops tension on the protruded disc. This exerts a retrusive effect on the disc.

It blends with the inferior stratum and with the joint capsule, and is highly vascular.

Attachment of the disc to muscle   On the medial and antero-medial sides the superior part of the lateral pterygoid muscle gains true tendinous insertion into the anterior extension and anterior band areas of the disc (Rees, 1954; Griffin and Sharpe, 1960; Moffett, 1984).

The mandibular capsular muscle gains insertion to the antero-lateral aspect (Koritzer and Suarez, 1983). Adjacent parts of the masseter, temporal and zygomatrico-mandibular muscles also have a direct connection with the antero-lateral aspect of the disc (Rees, 1954; Schumacher, 1961; Christensen, 1969; Widmalm, 1987).

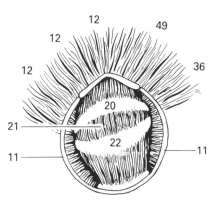

Figure 52 Capsule and disc of the craniomandibular joint viewed from above to show the muscle tissue attaching to them (diagrammatic): (11) joint capsule; (12) superior part of lateral pterygoid; (20) anterior band of disc; (21) intermediate zone of disc; (22) posterior band of disc; (49) zygomatico-mandibular muscle; (36) temporal muscle

Summary
Muscle tissue is inserted into the disc or capsule all the way around on the medial, anterior and lateral aspects.

Contraction of this muscle tissue can exert tension on the meniscus in different directions and of varying magnitude.

The disc attachment to the middle ear   A fibro-elastic extension of the superior stratum of the bilaminar zone passes from the glenoid fossa into the middle ear at the junction of the squamo-tympanic and petro-tympanic sutures in the postero-medial part of the fossa (Figure 38). This fibro-elastic structure is continuous with the anterior malleolar ligament which is attached to the anterior process of the malleus (Rees, 1954; Pinto, 1962). Burch (1966) has provided evidence that the anterior malleolar ligament is an upward continuation of the sphenomandibular ligament (p. 94; Figure 56). Moffett (1984) has traced the embryological development of the disc of the craniomandibular joint and lateral pterygoid in detail (see also p. 94).

Summary
A fine, tendinous extension of the bilaminar area passes through the squamo-tympanic suture into the middle ear. It may have functional significance.

Vascularization of the disc
The articular areas of the disc are densely collagenous and firm. They are avascular and without sensory receptors of any kind. The articular areas are surrounded by less dense fibrous areas that are well vascularized. The most vascular parts of the disc are found in the anterior extension and bilaminar zone (Griffin and Sharpe, 1960; Mahan, 1980).

With the mandible in the postural position, the bilaminar zone, in particular postero-superiorly to the condyle and antero-inferiorly to the squamo-tympanic fissure, is rich in blood vessels. The articular arteries are derived from the auriculotemporal, anterior tympanic and first part of the masseteric arteries. There are numerous veins present. Glomus-cell, arteriovenous shunts are also a feature of this area (Boyer, Williams and Strevens, 1964).

The anterior extension is also very vascular, especially inferiorly.

Around the sides of the articular part of the disc wide vascular channels, arteries and veins join these two major areas of vascularization (Boyer, Williams and Strevens, 1964).

Summary

1. The anterior extension and bilaminar zone are richly vascular.
2. Arteriovenous shunts are present.
3. The major anterior and posterior vascular areas are connected by vessels running on the medial and lateral extremities of the joint.

The capsule of the craniomandibular joint

The capsule of the joint is a loose, tapering, thin-walled cuff of white collagenous tissue enclosing the bony parts and disc of the joint. It has a wide attachment around the glenoid parts from which the supero-inferiorly-aligned collagen bundles taper to a much narrower attachment on the back and sides of the neck of the condylar process (Kraus, Jordan and Abrams, 1969; Figure 54).

The functions of the craniomandibular joint capsule are to:

1. Enclose and delineate two joint cavities.
2. Support the synovia and confine the synovial fluid.
3. Support a dense neural end-organ complex.
4. Provide continuity of attachment from muscle to disc.
5. Provide a weak restraining action.

Glenoid attachment
The joint capsule fasciculi are attached superiorly to the temporal bone just outside the circumductory pathway of the mandibular condyle–disc assembly (see pp. 72–74).

Posteriorly, the attachment is broad and includes the posterior articular lip, postglenoid process and whole anterior surface of the tympanic part of temporal bone.

Laterally the capsule is attached to the narrow crest of bone running between the postglenoid process and the lateral aspect of the articular tubercle.

Along the short medial boundary of the glenoid fossa the capsule attaches to the edge of the entoglenoid process, the medial glenoid plane and the medial crest of bone on the glenoid (articular) eminence.

Anteriorly, the limits of the circumductory path of the condyle on the preglenoid plane are unclear on many dried skulls. There is no clearly-defined joint capsule in this area. The limits of the joint are defined by the attachment of the anterior

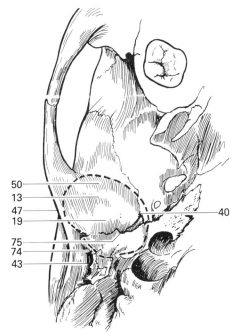

Figure 53 Attachment of the craniomandibular joint capsule to the temporal surface of the skull. This attachment is well clear of the circumductory path of the condyle–disc assemby: (13) crest of articular eminence; (19) glenoid fossa; (40) medial glenoid plane; (43) external auditory meatus; (47) tubercle of zygomatic arch; (50) preglenoid plane; (74) postglenoid process; (75) squamo-tympanic suture

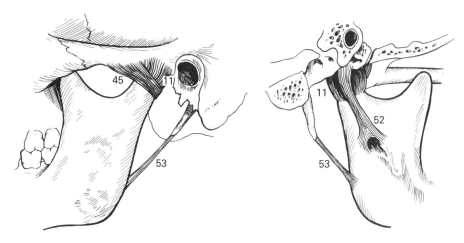

Figure 54 The capsule of the craniomandibular joint, with the temporomandibular, stylomandibular and sphenomandibular ligaments: (11) joint capsule; (45) temporomandibular ligament; (52) sphenomandibular ligament; (53) stylomandibular ligament

extension of the disc to the infratemporal skull surface (Figures 45 and 50). This is a weak attachment.

Posteriorly, the capsule and the inferior stratum of the bilaminar zone are not separable. The capsule fibres are distinguishable only by their situation on the outside and their direct course from the temporal bone to the mandible (Rees, 1954; Figures 47 and 50).

Anteriorly, the capsule is very poorly developed. Hence the upper and lower attachments of the anterior extension of the disc delineate the anterior boundaries of the joint cavities (Figures 45, 50 and 51).

The sides of the medial and lateral triangular flaps of the disc are attached to the capsule, making a medial and lateral recess or sulcus ('gutter'), the meniscocapsular sulcus, on each side in the upper joint compartment. This is evident in frontal section through the joint or in a superior approach to the upper joint cavity (Rees, 1954; Kraus, Jordan and Abrams, 1969; Mahan, 1980; Figures 49 and 51).

Condylar attachments
The capsule is attached to the condyle below the poles, medially and laterally, and to the lower border of the posterior joint surface of the condyle (Figures 45 and 49). The poorly-defined anterior capsule is attached to the inferior lip of the articular, facet, anteriorly (Figures 45 and 47).

The capsule on the medial and lateral parts of the joint is very slack and its attachment to the sides of the triangular flaps of the disc does not in any way hinder translatory movements of the disc (Figure 49 and 56).

The capsule, although composed of white collagen, is quite thin and much too slack to support the joint parts in function (Rees, 1954).

The lateral wall, especially anteriorly, supports a very dense, neural end-organ complex in the subsynovial and collagenous outer layer. The medial wall is innervated in a similar fashion, but the receptor complex is not as dense (Kawamura, 1974).

Summary

1. The capsule of the joint is composed of a thin layer of vertically-oriented, white collagen bundles.
2. It has the form of a slack cuff almost completely enclosing the bony areas of the joint.
3. It is attached to the disc forming upper and lower joint cavities.
4. It is lined by synovial membrane and supports a very dense neural end-organ complex.
5. It does not limit joint movement.
6. It is too weak to stabilize joint parts.

The ligaments of the craniomandibular joint

The capsule of the craniomandibular joint is quite thin and, on its own, is inadequate to support the joint parts in function or limit their movements.

The capsule is reinforced in its limiting function by special ligaments (Ärstad, 1954). These are the temporomandibular ligament, the collateral ligaments and the accessory ligaments.

The temporomandibular ligament

This is the main ligamentous reinforcement of the joint.

It is a well-developed, fan-shaped structure located on the lateral aspect of the joint (Figures 50 and 55).

The white collagenous fascicles arise along a wide band of the temporal bone, on the lateral side of the tubercle of its zygomatic process. The attachment is often marked by a raised, roughened ridge in the dried skull (Sicher, 1960). From their wide origin, the fascicles run obliquely, postero-inferiorly, to insert on the posterior of the neck of the condylar process of the mandible, gaining attachment below and behind the lateral pole of the condyle.

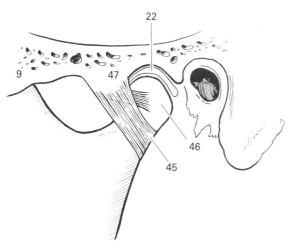

Figure 55 Diagrammatic representation of the lateral ligaments of the craniomandibular joint (after Ärstad, 1954). The capsule has been left out for clarity: (9) zygomatic arch; (22) posterior band of disc; (45) temporomandibular ligament; (46) lateral collateral ligament; (47) tubercle of zygomatic arch

The temporomandibular ligament is identifiable by its fascicle alignment and as a perceptable thickening of the lateral wall of the capsule. It is not separable from the lateral wall of the capsule by blunt dissection.

Like the lateral wall of the capsule, it is liberally endowed with neural receptor end-organs (Kraus, Jordan and Abrams, 1969). The ligament is vascularized (Bell, 1982).

Bilaterally, the temporomandibular ligaments form the main suspensory mechanism of the mandible, mechanically resisting gross disarticulation in a postero-inferior direction during functional movements.

When the condyle is fully seated in the fossa the anterior fibres of this ligament become taut helping to prevent further distal displacement of the mandible (Ärstad, 1954).

When the mandible is depressed the anterior fibres of the temporomandibular ligament become taut, as the posterior border of the mandible swings backwards. Tension on the ligament with further depression is one of the factors which compels the condyle to translate anteriorly as the mouth opens wider.

In young healthy joints the temporomandibular ligaments may restrict direct lateral movements and support the joint in functional positions (Kraus, Jordan and Abrams, 1969).

The collateral ligaments
On the inner sides of the joint capsule, horizontally-orientated, narrow bands of white collagenous tissue are located on the lateral and medial sides. On the lateral wall they run from the tubercle of the zygomatic arch to the lateral pole of the condyle (Figure 55) and constitute a definitive ligamentous structure (Ärstad, 1954; Sicher, 1960). On the medial side, a similar, but less well-definable, horizontal band of collagen fascicles runs to the medial pole from the medial side of the articular eminence (Ärstad, 1954; Kraus, Jordan and Abrams, 1969).

The collateral ligaments are innervated and vascularized.

The medial and lateral collateral ligaments become tense when the mandible is fully seated in the glenoid fossa and are believed to be, in conjunction with the temporomandibular ligaments, mechanical stabilizing factors in the establishment of a mandibular reference position in the horizontal plane (McCollum and Stuart, 1937).

The collateral ligaments biologically and mechanically restrict distal displacement of the condyles so that the retrocondylar tissues are not traumatized. The lateral collateral ligament probably becomes tense in the working-side joint as the working condyle pivots during a lateral mandibular movement. Owing to this, the vertical axis for lateral movement may be stationed close to the attachment of the lateral collateral ligament to the condyle towards the end of the pivoting movement (DuBrul, 1980).

Maximum intercuspation of the teeth places the mandible anterior to the position determined by the collateral ligaments. Loss of back teeth may result in stretching or disruption of these ligaments and so alter the biological and mechanical control of the posterior displacement of the condyles in the joint.

Other associations
The details of disc–condyle attachment have been dealt with above (pp. 83–85). The innermost zone of ligamentous tissue covers the poles of the condyle. With the

condyle–disc assembly in the normal postural position the collagen fibres are aligned mainly in a vertical direction.

Blended with the above fibres at the poles of the condyle, but running in a horizontal direction, are the collagen fibres of the collateral ligaments, while imperceptably blended with both are the mainly vertically-aligned fibres of the thin joint capsule (Figures 49 and 51). Reinforcing the lateral surfaces of the joints, the temporomandibular ligaments form perceptible thickenings with fibres aligned obliquely in a postero-inferior direction (Figures 51 and 54).

Innervation of the ligaments

The ligamentous structures and capsule of the craniomandibular joint are densely innervated for proprioceptive monitoring of jaw positions and movements (Kawamura et al., 1967). The muscles which move the mandible are the chief determinants of mandibular position and movements, and they accomplish this as efficiently and atraumatically as possible on the basis of information derived from the proprioceptors of the masticatory system. Those located in the joints make a major contribution to this guidance, the kinesthetic sensory input for jaw guidance. (Kawamura et al., 1967).

Stretching, tearing or disruption of the joint capsule or ligaments can interfere with the kinesthetic sensing of these tissues and hence, with accurate monitoring of mandibular movement (Bell, 1982; Kawamura, 1968).

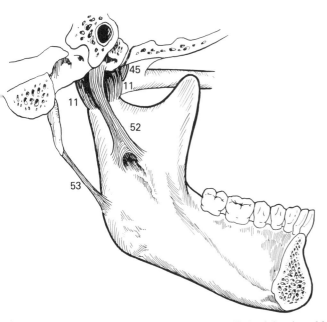

Figure 56 Ligaments and capsule of the craniomandibular joint viewed from the medial side: (11) joint capsule; (45) part of temporomandibular ligament; (52) sphenomandibular ligament; (53) stylomandibular ligament

Accessory ligaments

Two ligamentous structures were formerly described as accessory ligaments of the craniomandibular joint, they are:

1. The sphenomandibular ligament. This is attached between the spine of the sphenoid and the medial surface of the mandible in the region of the lingula. It is a thin membrane with poorly-defined borders. It has surgical significance but no known functional task to perform (Figure 56).

   The sphenomandibular ligament attachment to the spine of the sphenoid bone is disputed by Burch (1966). Burch demonstrated that in the majority of cases the upper attachment of this ligament was to the petro-tympanic fissure (Figures 38 and 53, pp. 72 and 89) and through it to the malleus of the middle ear. Tension on the sphenomandibular ligament moved the malleus. The extension of the sphenomandibular ligament passing below the corda tympani nerve into the middle ear cavity becomes the anterior malleolar ligament (Burch, 1966).

   This extension of the sphenomandibular ligament is often closely related and joined to the fibro-elastic structure passing through the squamo-tympanic suture from the superior stratum of the bilaminar zone to the malleus of the middle ear (Rees, 1954; p. 88).

2. The stylomandibular ligament. This is a collagenous reinforcement of local fascia extending from the styloid process to the posterior border of the angle of the mandible. It is an identifiable structure of considerable surgical significance. Functionally, it may restrict extreme ipsilateral protrusive movements of the mandible. Rotatory movements will approximate the attachments of this ligament (Figure 56).

Summary

1. The joint capsule is reinforced by local ligamentous structures, laterally and medially.
2. On the inner surface of the capsule the strengthening fascicles are aligned horizontally with the mandible in the maximum intercusping position. They attach to the poles of the condyle and are designed to restrict posterior movement.
3. On the external, lateral surfaces of the joints the strengthening fascicle alignment is obliquely down and backwards and the attachment is to the posterior of the neck of the condylar process.

   This temporomandibular ligament restricts distal and inferior movements of the mandible and supports functional joint positions.
4. The ligamentous supporting structures are liberally endowed with pain and proprioceptive neural end-organs, and are vascularized. They constitute a major jaw guidance factor, kinesthetically and mechanically. When subjected to excessive stress they may cause pain and become inflamed.
5. The accessory ligaments have little known functional significance.

The synovial membranes of the craniomandibular joint

The articular surfaces of the joint are lubricated and nourished by synovial fluid which is secreted into the joint compartments by the synovial membranes.

The synovial membranes line the inner surface of the capsule and are reflected from the capsule onto the bony parts of the joint up to the edges of their articular surfaces. They are also reflected onto the non-articular edges of the discs.

The synovial membranes do not extend onto the articular surfaces. They are present all around the sides of the articular, weight-bearing areas, but are especially conspicuous in the anterior and posterior of the lower compartment and in the posterior and menisco-capsular sulcus of the upper joint compartment (Figures 45 and 52). The membranes, when the disc is fully seated in the fossa, are usually very lax and folded upon themselves to give a villous appearance (Rees, 1954; Griffin and Sharpe, 1960; Mahan, 1980).

The synovial membranes consist of a layer of specialized secretory cells on the surface. These cells resemble fibroblasts and the layer is usually three to four cell layers thick.

The layer of synovial cells is closely related to highly vascularized, subsynovial tissue in which there is a rich plexus of arterial and venous capillaries and many, lymphatic channels. The arterioles and venules have relatively thick walls (Griffin and Sharpe, 1960).

No neural end-organs are found in the synovial membrane except those on the blood vessels walls. The membrane is insensitive to pain stimuli (Bell, 1982).

The synovial membrane and the subsynovial zone is well supported by the capsule of the joint and the bilaminar tissue. There is no well-defined capsule around the anterior of the joint. The synovial membrane of the anterior part of the upper joint cavity is supported only by the fibrous areolar tissue which joins the anterior extension of the disc to the preglenoid plane (Figures 45 and 52).

The synovial fluid is sparsely secreted and enough is present to act as a highly efficient lubricant and as a buffer against compression. The completely avascular articular surfaces of the eminence, condyle and disc are supplied with essential metabolites by the synovial fluid. The fluid also cleanses worn pieces and cell catabolites off the joint surfaces. The synovial fluid probably is circulated passively by movement of the surfaces over each other and pressure developed between them within the joint compartments (MacConail, 1950).

The exact mechanism of lubrication of the craniomandibular joint is not known (DuBrul and Menekratis 1981). Since there is no normally structured cartilage in the joint a 'weeping' lubrication (MacConail, 1950) is not possible. Thus it is likely that the mechanism is a modification of 'boundary lubrication'. In the latter, the secreted fluid reacts chemically with the articular tisue to form a greasy, slippery film which reduces friction between the rubbing surfaces. The forces developed in the craniomandibular joint probably involve great tensile stress, hence the fibrous nature of the articular surface (DuBrul and Menekratis, 1981; see also 'joints in general' pp. 66–67).

Summary

1. The synovia are specialized secretory membranes which provide a nutrient, lubricating and cleansing service for the avascular, load-bearing joint surfaces.
2. The synovia are not innervated.
3. The mechanism by which the craniomandibular joint is lubricated is not known with certainty.

Vascular and nerve supply to the craniomandibular joint

Gross supply

Every named artery within 30 mm gives one or two branches to the joint (Williams and Warwick, 1980).

Articular branches enter the capsule from the:

1. Superficial temporal artery (posteriorly).
2. Maxillary artery (posteriorly).
3. Masseteric artery (anteriorly).
4. Intramuscular vascular bed of external pterygoid muscle.
5. Periosteal blood vessels all around the craniomandibular joint.
6. Within the mandible and/or temporal bone and enter the capsule along its attachment to either bone.

Capsular plexus

Within the capsule an extensive vascular plexus is formed, fed by the gross supply from all directions. The capsular plexus extends towards the articular areas of the joint, all around its circumference, through the capsule–disc union. The vessels divide into smaller components and are simple capillary loops at the margins of the weight-bearing articular areas and the synovial folds. Thus the vessel size becomes smaller and the number of vessels becomes greater as the articular surfaces are approached (Griffin, 1959; Boyer, Williams and Stevens, 1964).

The venous drainage is along similar routes to the arterial supply outlined above. A rich plexus of veins characterizes the anterior extension and bilaminar zone in the disc. Arteriovenous shunts are common in these areas (Griffin and Sharpe, 1960; Boyer, Williams and Strevens, 1964).

Nerve supply

The predominant source of innervation is via articular branches of the auriculo-temporal nerve which is located posterior to the joint. A lesser supply of nerves is derived from the articular branches of the masseteric nerve and the deep temporal nerves. The latter nerves pass mainly to the front of the joint (Williams and Warwick, 1980).

Chapter 10

The retrocondylar tissues

Owing to non-standardized terminology the retrocondylar tissues are sometimes confused with the bilaminar zone tissues.

The retrocondylar tissues are those situated posterior to the condylar process of the mandible. They are comprised of:

1. The bilaminar zone of the disc.
2. The part of the joint capsule posterior to the condyle.
3. Vascular elements.
4. Neural elements.
5. Dense fibrous connective tissue encasing the total contents.

They form a bulk of tissue 6–7 mm thick (Koritzer and Kenyon, 1983). Note that points (1) and (2) have been described in detail above (see pp. 81 and 90).

The major blood vessel is the maxillary artery which is 3–4 mm in diameter (Koritzer and Kenyon, 1983) and divides into the deep auricular artery, the anterior tympanic artery and the middle meningeal artery (± the accessory meningeal artery).

The only significant venous concomitant is 2–3 mm in diameter. The deep auricular artery gives off an articular branch to the craniomandibular joint. This is the only terminal artery in the area.

The blood vessels are encased in dense fibrous connective tissue which is continuous, laterally, with the temporal periosteum and its associated muscle tissue and, medially, with the infratemporal periosteum and the superior part of the lateral pterygoid muscle.

The posterior temporal and superior parts of lateral pterygoid muscles probably have the potential to transmit tension to the retrocondylar blood vessels via the fibrous encasing tissue (Koritzer and Kenyon, 1983).

It is unlikely that the distension of these blood vessels could occur to an extent that would result in effects in the ears or the nearby cranial cavities.

When translatory movements occur in the craniomandibular joints the space formerly occupied by the translating parts becomes occupied (Rees, 1954) by adjacent mobile or deformable tissues which can be 'pressed' into the potential space by atmospheric pressure or by movement of blood into and out of the extensive vessel network of the anterior extension and the bilaminar zone. The latter are interconnected around the sides of the articular parts of the joint. When the condyle–disc assembly translates forwards the bilaminar vascular network rapidly fills with blood and the anterior extension network is emptied. The reverse occurs when the condyle is returned to the glenoid fossa (Figure 47, p. 83).

The retrocondylar tissues are no longer believed to act as an 'erectile bumper' to prevent posterior displacement of the condyle–disc assembly, as proposed by Zenker (1956). Distal displacement of the condyle–disc assembly is prevented by the joint ligaments and the lateral pterygoid muscle (Reese, 1954; Sicher, 1960).

When distal displacement of the condyle is not prevented, then damaging compression of the retrocondylar tissues can occur (Steinhardt, 1958; Griffin and Sharpe, 1960).

Chapter 11

Functional movements and relationships of the craniomandibular joint parts

Rees (1954) described the functional movements within the craniomandibular joints and the changing relationships of the joint parts throughout those movements (Figure 58).

In healthy well-structured joints in the inferior joint compartment the condyles can rotate (or spin) around a horizontal axis (Figures 57, 58, 61 and 62). This is a bilateral movement occurring simultaneously in both joints. It is part of the simple hinge movement of the mandible in the vertical plane (Meyer, 1865; Figure 58).

As indicated above (pp. 83–85), the scope for lateral or twisting movements between the condyle and disc in the lower compartment is due to the inherent elasticity of the parts only (Kraus, Jordan and Abrams, 1969; Mahan, 1980). This is

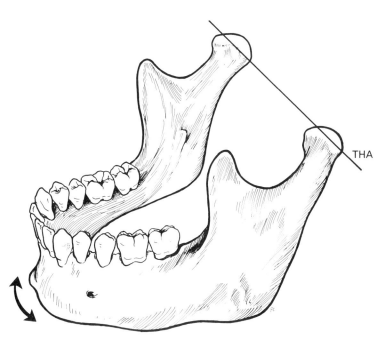

Figure 57 Rotation in the sagittal plane transverse horizontal axis (THA), passing through both condyles

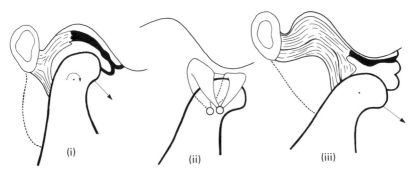

(i)   (ii)   (iii)

Figure 58 Rotation in the inferior joint compartments (Rees, 1954). This can occur with the discs in the glenoid fossae, (i) or with the condyle–disc assembly translated, (ii) and (iii)

because the condyle–disc parts are very closely fitted and tightly bound, permitting rotatory movement only (Mahan, 1980; Bell, 1982).

In healthy well-structured joints in the superior joint compartment, while the condyle–disc assembly is a firmly-bound and closely-adapted mechanism, the joint capsule is attached to both in such a way that separate superior and inferior joint compartments are formed (Figures 43–45). The capsule is quite slack and does not in any way hinder the normal, functional, surface-contact, sliding movement of the condyle–disc assembly in the upper joint compartment (Meyer, 1865; Reese, 1954).

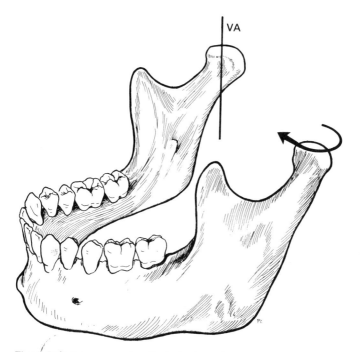

Figure 59 Rotatory movement in the horizontal plane vertical axis (VA) in the region of the working-side condyle. Sagittal perspective

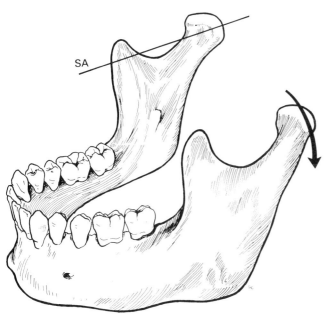

Figure 60 Rotation in the frontal plane, around an axis in the sagittal plane, through the working-side condyle. Sagittal axis (SA). Sagittal perspective

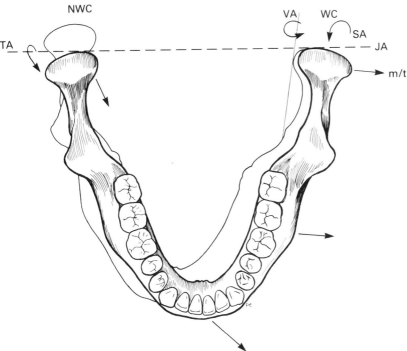

Figure 61 Left lateral movement with mandibular lateral translation: (NWC) non-working-side condyle; (WC) working-side condyle; (TA) transverse hinging axis; (VA) vertical axis; (SA) sagittal axis; (m/t) mandibular lateral translation. Perspective from above the jaw

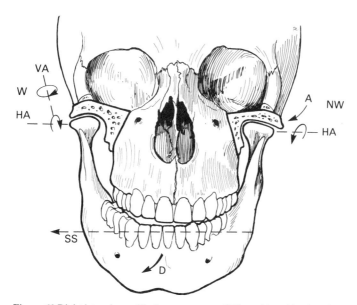

Figure 62 Right lateral mandibular movement: (W) working side; (NW) non-working side; (A) antero-medial swerve of non-working-side condyle rotates the mandible around the vertical and sagittal axes on the working side. Depression of the body of the mandible (D), in the sagittal plane, to clear the vertical overlap of the teeth, causes rotation around the hinging axis (HA) passing transversely in the region of both condyles: (VA) vertical axis; (m/t) mandibular lateral translation

Thus the condyle–disc assembly can slide to and fro on the articular surface of the eminence (Figure 58), pivot and rotate around a vertical axis (Figures 59 and 61) and slide medio-laterally and vice versa (Figures 61 and 62).

Owing to the laxity of the joint capsule combinations, different movements can occur simultaneously in the superior compartments of the right and left joints (Figures 61–63) and the condyle–disc assembly can be distracted away from the eminence.

Lateral movement of the jaw

To make a sidewards movement with the mandible, the condyle–disc assembly, on the side towards which the jaw is moving, does not advance anteriorly very much; this side pivots (Meyer, 1865; Figures 59, 61 and 62). The condyle–disc assembly can turn in the horizontal plane because of the slack nature of the capsule. The most stable and easiest place for the pivoting to occur is against the postero-superior surface of the eminence (Figures 58i, 59, 61 and 62).

During a lateral movement, on the side away from which the mandible is moving, the condyle–disc assembly progresses antero-inferiorly and medially on the eminence, i.e. the translation is marked compared with the opposite side (Figures 61 and 62).

During a lateral movement, in addition to the pivoting on one side and translation on the other, the whole jaw can translate laterally towards the side to which it is moving, the mandibular lateral translation (Ferrein, 1774; Figures 61 and 62). The latter is a very small component of movement, severely limited by the joint anatomy, and brought about by the pterygoid muscles on the side of the protruding condyle.

All of the above components of a lateral movement of the jaw, pivoting, anterior sliding and lateral sliding, occur in the upper joint compartments. Simultaneously, in the lower joint compartment, a simple, hinge movement of varying extent occurs.

During the movement:

1. The tight binding of the medial and lateral discal ligaments, the close fit of the disc on the condyle and the wedge-like thickening of the disc laterally, medially, anteriorly and posteriorly maintains the correct relationship of these parts.
2. The activity of the elevator muscles maintains the condyle–disc assembly in contact with the articular eminence (Figure 63).

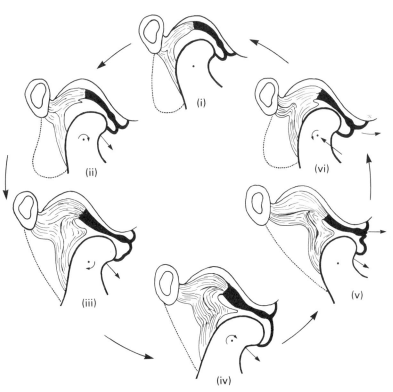

Figure 63 Structural relationships of the craniomandibular joint during functional jaw movement (after Rees, 1954). The sequence (anticlockwise) shows: (i) condyle–disc assembly with disc fully seated in fossa; (ii) condyle rotating on disc as mandible is depressed; (iii) condyle continues to rotate and condyle–disc assembly translates anteriorly–tissues move into space created posterior to condyle; (iv) mandible depressed – disc close to crest of condyle; (v) and (vi) disc moving and condyle rotating in reverse directions to (iii) and (ii)

Rotation in the sagittal plane occurs in the lower joint compartment. This is a simple hinge-like movement and, with the discs stationed in the glenoid fossae, can occur only over about one-half of the range of mandibular depression. It is called the centric jaw relation arcing movement (Figures 57 and 58). At the caudal limit of this movement, the tautness of the temporomandibular ligaments and compression of structures dorsal to the mandible compel anterior translation of the condyle if further depression is to be achieved.

Hinging movement in the lower compartment can and does occur when the condyle–disc assembly translates. The axis for the hinging movement moves with the mandibular condyles (Meyer, 1865).

Relationship of craniomandibular joint parts during mid-sagittal depression of the mandible (after Rees, 1954)

Starting with the disc fully seated in the glenoid fossae the mandible can hinge up and down over a range of approximately 20–25 mm (Figures 57 and 63i and ii). At 25 mm depression, tautness of the temporomandibular ligaments compels anterior translation of the condyle–disc assemblies (Figure 63iii and iv), during which further rotation in the lower compartment occurs.

As the condyle–disc assembly translates, the bilaminar tissues, capsule and skin are sucked into the space vacated by them, the blood spaces of the bilaminar tissues fill with blood and the superior stratum of bilaminar zone is passively stretched (Figure 63iii and iv).

The disc changes shape as required to adapt itself to the changing shape of the eminence and the pull of the bilaminar tissue (Rees, 1954; Bell, 1982).

In the most depressed position of the jaw, the anterior band of the disc is out on the preglenoid plane and the crest of the condyle is anterior to the crest of the eminence. This is an unusual functional position. It may occur as a result of a wide open mouth yawn.

During rotation or translation of the condyle–disc assembly, as the mandible is depressed, the disc appears to rotate posteriorly on the condyle so that the posterior band of the disc and posterior attachment of the capsule on the neck of condylar process approach each other and the anterior attachment of the disc to condyle moves further away from the disc (Figure 63ii, iii and iv).

Chapter 12

The adaptation of structure to function in the craniomandibular joint

To maximize efficiency, space utilization and overall streamlining in the masticatory system, a third class lever system is required. Hence the condylar processes are ideally positioned at the top and back of the rami, 20 to 30 mm above the level of the occlusal plane. The craniomandibular joints provide the environment in which the condylar processes can act as:

1. Stabilizers of mandibular positions and movement.
2. Fulcrums and pivots for mandibular movements, and the use of the mandible as a lever.
3. Guides to mandibular movements.
4. Attachments for the mandible to the skull.

All of these functions have to be fulfilled if the mandible is to move precisely enough to articulate the teeth satisfactorily, apply enough force to masticate food or provide a base from which to elevate the hyoid bone in swallowing.

Two joints are required in order to:

1. Perform the same movements on right and left sides – symmetrical movements, as in the ingestion of food or drink.
2. Stabilize the U-shaped dental arch or body of the mandible in symmetrical or asymmetrical movements and positions.
3. Enable three-dimensional movements of the mandible.

Positioning the joints, bilaterally on the posterior of the mandible and anteriorly on the dental arch, means that the mandible acts as a third class lever. This is the least efficient form of lever, but it enables the tissues of the masticatory system to be arranged in the most compact and streamlined fashion and still perform all the functions required of it with ease and smoothness.

The human mandible is characterized by the three-dimensional nature of its movements. To carry out such movements both joints must be able to perform rotatory and linear movements. Furthermore, each joint must be able to accommodate both types of movement simultaneously. In these circumstances the bony parts of the joints can not be made to fit tightly into each other. In the upper compartment of the human jaw the parts are a loose, approximate fit within a loosely-structured joint capsule. Articular contact is achieved indirectly, through the movable, flexible articular disc which can ensure the type of articular contact required by the conditions which exist at any one time.

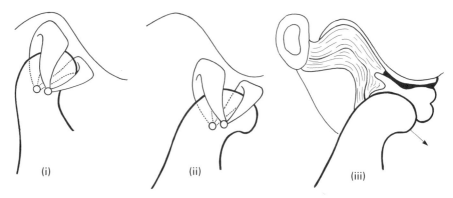

(i)   (ii)   (iii)

Figure 64 During functional movements of the condyle the disc–condyle relationship changes. In either rotation or forward translation of the condyle the condyle rotates forwards relative to the disc or vice versa (i), (ii). The posterior attachment of the disc to the condyle rotates upwards towards the disc (ii). In addition, during translation, the stretched bilaminar tissues pull distally on the disc and may change its shape temporarily – adapting it to the irregular outline of the joint surfaces (iii)

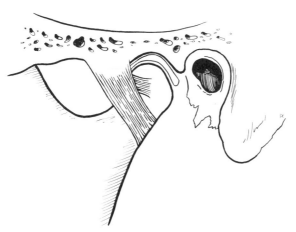

Figure 65 The temporomandibular and collateral ligaments prevent distal and postero-inferior displacement of the condyle during normal function (after Árstad, 1954)

Articular function

The articular surfaces of the craniomandibular joint consists of an avascular, dense, fibrous tissue. This tissue can resist the shearing, compressive force applied to it.

There is a cartilaginous-like matrix in the deep surface layers and occasional cartilage cells may be seen in older groups (Mahan, 1980; Moffett, 1984). The regenerative ability of the fibrous tissue is greater than cartilage. This makes it more suitable for the craniomandibular joint where the articular surfaces may be subjected to much micro-trauma, and where the demand for adaptive remodelling is high (Moffett, 1980).

The lower craniomandibular joint compartment forms a tightly-structured hinge joint (ginglymus). The tightly fitting disc permits movement of the condyle, in relation to it, in one plane only.

The condyle–disc assembly forms the lower component in the upper joint compartment. It slides on the temporal articular facet. The upper joint is designed to allow a multidirectional sliding movement. To enable this the surfaces are nearly flat and the upper joint sliding is unrestricted by ligamentous binding within 'the circumductory path' of the joint (DuBrul, 1980). The condyle–disc assembly can slide anywhere on the temporal facet – the articular eminence, the crest of the eminence and the preglenoid plane.

Contact between the condyle–disc assembly and the temporal articular facet is maintained by the pull of the elevator muscles across the joint surfaces in an antero-superior direction (Chapter 6, pp. 52–54).

Postero-inferior displacement of the condyle–disc assembly is prevented by the muscle tissue which elevates and protrudes the mandible. The temporomandibular ligaments resist functional postero-inferior displacement and its proprioceptive sensors probably have a major input to the pattern generator concerned with preventing this displacement (Figures 65 and 66).

Excessive distal displacement is prevented by the lateral pterygoid muscle and resisted by the collateral ligaments and their sensory input to the central nervous system (Figure 65).

Roles of the disc

Sharp contact between the joint surfaces is a requirement for stability in a synovial joint (Williams and Warwick, 1980). This is achieved in the various static and dynamic relationships of the craniomandibular joint by the disc and its positioners – the muscle tissue attached to the cranio-mandibular joint and the superior stratum of the bilaminar zone tissues.

When the condyle–disc assembly is forcefully squeezed against the eminence, the intermediate zone of the disc is interposed between the condyle and the eminence and is rotated into position on the head of the condyle by the applied force and its wedging, self-centering morphology (Figures 42–44; Moffett, 1984).

With the mandible in the postural position the condyle–disc assembly drops away slightly from the upper surface of the upper joint compartment (Nevakari, 1956). The postural activity of the superior part of the lateral pterygoid muscle is sufficient to rotate the disc anteriorly on the condyle, since in the postural position the elastic tissue of the superior stratum of the bilaminar zone is resting. The disc is rotated anteriorly until the thick posterior band provides the stabilizing contact for the postural position (Bell, 1982).

During sliding movements in the joint the condyle–disc assembly moves away from the glenoid fossa, sliding down the eminence. In the process, the superior stratum of bilaminar zone is stretched and exerts elastic tension on the disc. This rotates the disc posteriorly on the condyle. This effect is increased by any rotation which takes place in the lower compartment of the joints (Figures 48 and 64; Rees, 1954).

The elastic pull on the translating disc has the effect of adapting it to the temporal articular surface – eminence, eminence crest, and preglenoid plane (Figures 48, 63 and 67).

108

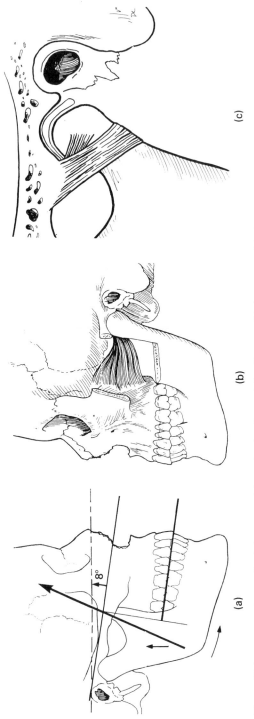

(a)

(b)

(c)

Figure 66 Antero-superior result of elevator activity maintains the condyle–disc assembly in contact with the postero-superior surface of the articular eminence (a). Postero-inferior displacement of the condyle–disc assembly is prevented by the lateral pterygoid inferior head and the elevators (b and c). Functional postero-inferior displacement is prevented by the temporomandibular and collateral ligaments. The rich proprioceptive innervation of these ligaments and the craniomandibular joint capsule have a major role in programming the pattern generator for mandibular movements

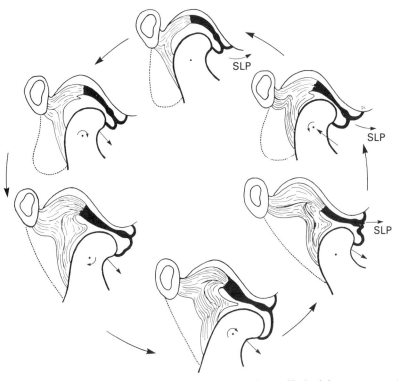

Figure 67 Sagittal view of structural relations during craniomandibular joint movements (after Rees, 1954): (SLP) pull of superior head of lateral pterygoid muscle

During retrusion to the glenoid fossa the elastic traction of the superior stratum gradually decreases. During retrusion under load, the muscle tissue inserted into the joint capsule – assembly (e.g. the superior part of lateral pterygoid, the zygomatico-mandibular muscle, the mandibular capsular muscle and the temporalis) becomes active and controls the velocity of retrusion (Figures 66 and 67).

Thus the anterior-posterior translation and rotation of the disc is capable of adaptation to the conditions of the time so that the jaw is stabilized during its movements and in the positions it takes up.

Chapter 13

Anatomical variation in the craniomandibular joint

Considerable variations occur in the size and form of the craniomandibular joint and in the adaptation of form to functional demands.

Size of the craniomandibular joint

In general, joint size is directly proportional to general skeletal size. Larger joints are present in larger subjects and vice versa. Joint part sizes tend to conform to each other also; larger condyles go with larger glenoid structures and disc, etc.

Significant discrepancies in joint part sizes usually result from developmental or growth anomalies. No data are available to indicate whether hereditary factors may give rise to structural anomalies of significance, e.g.:

1. Small condyles in large fossae.
2. Tight, effective attachment or adaptation of the disc to the condyle with loose attachment to the temporal bone or vice versa.
3. Strong, effective joint ligaments.
4. Proportional relationships between the disc and the bony parts which favour good mechanical operation, e.g. a well-developed posterior band in a deeply-vaulted fossa with a steeply-inclined glenoid eminence, as distinct from a poorly-developed posterior band. The latter would aid anterior displacement of the disc and posterior displacement of the condyle.

Form of the joint

The specific joint morphology has considerable functional, clinical and radiological significance.

Great variation in form does occur and this is well documented (Meyer, 1865; Lindbloom, 1960). The same may be true for the disc, but is not documented. In general, morphology of the condyle and glenoid structures corresponds, i.e. a condyle that is steeply convex in the frontal plane will usually be associated with a steeply concave fossa and eminence, while a condyle with a low cresendic form will be associated with a shallow fossa. However, these statements are not always valid. As pointed out previously, the condyle is a loose, rough fit to the glenoid structures. Anomalies of fit may be compensated for by size and form of the disc.

Adaptation of form to functional demands

The craniomandibular joint is a weight-bearing joint. The fibrous covering of the articular bone is believed to respond to increased stress by proliferation. This causes a local thickening of the articular surface (Hansson and Nordstrom, 1977; Bean, Omnell and Öberg, 1977; Moffett, 1984). The soft tissue response is accompanied by changes in the underlying bone such that its bony contour is changed (Hansson and Öberg, 1977). These local changes in the form of the joint occur almost exclusively in the lateral half of the joint, because that side of the joint is subject to the greatest stress. The changes are called 'deformations in form'.

Deformations in form have a negative effect on the disc, resulting in a thinning of the intermediate zone at its juxtaposed part (Öberg, Carlsson and Fajers, 1971). Perforation may eventually occur, although it is unusual below 40 years of age (Hansson and Öberg, 1977; Öberg, Carlsson and Fajers, 1971). Deformation in form begins to appear once the joint is full grown and developed. The incidence increases with age and it is common in joints over 40 years of age (Hansson *et al.*, 1979).

Deformations in form are associated with surface changes in the joints and so may be responsible for changes in joint sounds. They may eventually cause interference with joint function, especially in older age groups or when the disc becomes perforated.

References

Ärstad, T. (1954) *The Capsular Ligaments of the Temporomandibular Joint and Retrusion Facets of the Dentition in Relationship to Mandibular Movements,* Akademisk Forlag, Oslo

Balkwill, F. H. (1985) Best form and arrangement of artifical teeth for mastication. *Odontological Society of Great Britain Translations,* 5 (133), 1865–1867

Bean, L. R., Omnell, K. A. and Öberg, T. (1977) Comparisons between radiologic observations and macroscopic tissue changes in temporomandibular joints. *Dentomaxillofacial Radiology,* 6, 90

Bell, W. F. (1982) *Clinical Management of Temporomandibular Disorders,* Year Book Publishers Inc., New York

Boyer, C. C. Williams, T. W. and Strevens, F. H. (1964) Blood supply of the temporomandibular joint. *Journal of Dental Research,* 43, 224

Burch, J. G. (1966) The cranial attachment of the sphenomandibular ligament. *Anatomical Reconstruction,* 156, 433

Carlsöö, S. (1956) An electromyographic study of the activity and an anatomic analysis of the mechanics of the lateral pterygoid muscle. *Acta Anatomica,* 26, 339

Christensen, F. G. (1969) Some anatomical concepts associated with the temporomandibular joint. *Annals of the Australian College of Dental Surgery,* 2, 39

Cunningham, D. J. (1902) *Textbook of Anatomy,* 1st edn, Young J. Pentland, Edinburgh and London

Da Vinci, Leonardo (1452–1519) Italian artist, anatomist, inventor, architect, musician, scientist. First accurate representtion of the skull and masticatory system. Earliest description of the dental occlusion.

Dawson, P. E. (1974) *Evaluation Diagnosis and Treatment of Occlusal Problems.* C. V. Mosby & Co., St Louis

DuBrul, E. L. (1980) *Sicher's Oral Anatomy,* 7th edn, C. V. Mosby & Co., St Louis

DuBrul, E. L. and Menekratis, A. (1981) *The Physiology of Oral Reconstruction,* 1st edn, Quintessence International, Berlin

Ebert, H. (1939) Morphologische und funktionelle analyse des musculus masseter. *Zeitschrift für Anatomie und Entwicklungsgeschickte,* 109, 790

Ekholm, A. and Sürila, H. S. (1960) An electromyographic study of the function of the lateral pterygoid muscle. *Suomi Hammaslaak Toim,* 58, 90 (quoted by McNamara, 1973)

Ferrein, M. (1774) Sur les mouvements de las machoire inferieure. *Memories de L'Academic Royale des Sciences,* 427–448

Findlay, I. A. (1964) The movements of the mandibular inter-articular disc. *Proceedings of the Royal Society of Medicine,* 58, 671

Fischer, R. (1935) Die Öffnungsbewegungen des unterkiefers und ihre Wiedergabe am Artikulator. *Schweizerische Monatschrift für Zahnheilkunde,* 45, 867

Goodson, J. M. and Johansen, E. (1975) Analysis of human mandibular movement. In *Monographs in Oral Science,* Vol. 5, S. Karger, Basel

Grant, P. G. (1973a) Lateral pterygoid: Two muscles. *American Journal of Anatomy,* 138, 1

Grant, P. G. (1973b) Biomechanical significance of the instantaneous center of rotation: the human temporomandibular joint. *Journal of Biomechanics,* 6, 109

Gray, H. (1858) *Anatomy of the Human Body: Descriptive and Applied,* Longmans, London

Griffin, C. J. (1959) Mechanism of the blood supply to the synovial membrane of the temporomandibular joint. *Australian Dental Journal,* **4**, 379

Griffin, C. J. and Munro, R. R. (1969) Electromyography of the jaw closing muscles in the open-close-clench cycle in man. *Archives of Oral Biology,* **14**, 141

Griffin, C. J. and Sharpe, C. J. (1960) Structure of the adult human temporomandibular meniscus. *Australian Dental Journal,* **5**, 190

Hansson, T. and Nordstrom, B. (1977) Thickness of soft tissue layers and articular disc in temporomandibular joints with deviations in form. *Acta Odontologica Scand.,* **35**, 281

Hanson, T. and Öberg, T. (1977) Arthrosis and deviation in form in the temporomandibular joint. A macroscopic study on human autopsy material. *Acta Odontologica Scand.,* **35**, 167

Hannson, T., Öberg, T., Solberg, W. K. and Penn, M. (1979) Anatomic study of the TMJ's of young adults. Pilot investigation. *Journal of Prosthetic Dentistry,* **41**, 556

Henke, W. (1863) *Handbuch der Anatomie und Mechanik der Gelenke mit Rucksicht auf Luxation und Kontraktionen,* Leipzig und Heidelberg (quoted by Schumacher, 1961)

Hickey, J. C., Stacey, R. W. and Rinear, L. (1957) Electromyographic studies of mandibular muscles in basic jaw movements. *Journal of Prosthetic Dentistry,* **7**, 565

Hildebrand, G. Y. (1931) Studies in the movements of the human lower jaw. *Skandinavisches Archive für Physiologie* (Supplement zum 61 band), Walter De Gruyten & Co., Berlin U. Leipzig

Honée, G. L. J. M. (1972) The anatomy of the lateral pterygoid muscle. *Acta Morphologica Neerl – Scand.,* **10**, 331

Hrycyshn, A. W. and Basmajian, J. V. (1972) Electromyography of the oral stage of swallowing in man. *American Journal of Anatomy,* **133**, 330

Juniper, R. P. (1981) The superior pterygoid muscle. *British Journal of Oral Surgery,* **19**, 121

Juniper, R. P. (1984) Temporomandibular joint dysfunction: a theory based upon electromyographic studies of the lateral ptergyoid muscle. *British Journal of Oral and Maxillary Surgery,* **22**, 1

Kawamura, Y. (1968) Mandibular movement. In *Facial Pain and Mandibular Dysfunction* (eds. L. Schwartz and C. M. Chayes), W. B. Saunders & Co., Eastbourne

Kawamura, Y. (1974) Neurogenesis of mastication. *Frontiers of Oral Phyisology,* **1**, 77

Kawamura, Y., Majima, T., Kato, I. (1967) Physiologic role of deep mechanoreceptors in temporomandibular joint capsule. *Journal of the Osaka University Dental School,* **7**, 63

Kawamura, Y. and Kimura, T. (1971) Analysis of masticatory muscle function in the rabbit. *Journal of Dental Research,* **50**, 1148

Koole, P., *et al.* (1984) *Journal of Craniomandibular Practice,* **2**, 326

Koritzer, R. (1983) Georgetown Universtiy Dental School (personal communication)

Koritzer, R. T. and Kenyon (1983) Further description of the temporal muscle and retrocondylar temporomandibular area. (unpublished text)

Koritzer, R. T. and Suarez, F. (1980) Accessory medial pterygoid muscle. *Acta Anatomica,* **107**, 467

Kraus, B. S., Jordan, R. E. and Abrams, L. (1969) *Dental Anatomy and Occlusion,* Williams & Wilkins, Baltimore and London

Krogh-Poulsen, W. and Molhave, A. (1957) *Tandlaegebladet,* **61**, 265

Lindbloom, G. (1960) Anatomy and Function of the temporomandibular joint. *Acta Odontologica Scand.,* **17**, 7 (Suppl. 18)

Lipke, D. P., Gay, T., Gross, B. D. and Yaeger, J. A. (1977) An electromyographic study of the human lateral ptergyoid muscle. *American Association of Dental Research Abstracts,* 713

Lord, F. P. and Hanover, M. D. (1937) Movements of the jaw and how they are effected. *International Journal of Orthodontics,* **23**, 577

Luce (1889) *Boston Medical and Surgical Journal,* July 4th 1889 (Quoted by Walker, W. E. *Dental Cosmos,* January 1896)

Lundeen, H. C. and Gibbs, B. H. (eds.) (1982) Applied clinical research in form and functional relationships. In *Advances in Occlusion,* Wright, Bristol

McCollum, B. B. and Stuart, C. E. (1935) *A Research Report,* C. E. Stuart Gnathological, Ventura, California

McConail, M. A. (1950) The movements of bones and joints: The synovial fluid and its assistants. *Journal of Bone and Joint Surgery,* **32B**, 244

McConail, M. A. (1975) The muscular slings of the mandible. *Journal of the Irish Dental Association*, **21**, 57

McConail, M. A. and Basmajian, J. V. (1969) *Muscles and Movements, a Basis for Human Kinesiology*, Williams, Wilkins & Co., London

McMinn, R. M. H., Hutchings, R. T. and Logan, B. M. (1981) *Colour Atlas of Head and Neck Anatomy*, 1st edn, Wolfe Medical Publications, London

McNamara, J. A. (1973) The independent functions of the two heads of the lateral pterygoid muscle. *American Journal of Anatomy*, **138**, 197

Mahan, P. E. (1980) Temporomandibular joint in function and pathofunction. In *Temporomandibular Joint Problems: Biologic Diagnosis and Treatment*, (eds W. K. Solberg, and G. T. Clark), Quintessence Publishing Co., Chicago

Meyer, H. (1865) Das Kiefergelenk. *Archivs für Anatomie und Physiks*, 719

Mills, J. R. E. (1967) A comparison of lateral jaw movements in some mammals from wear facets on the teeth. *Archives of Oral Biology*, **12**, 645

Moffett, B. (1974) The temporomandibular joint. In *Complete Denture Prosthodontics*, J. J. Sharry (ed), McGraw Hill, New York

Moffett, B. (1980) Clinical biology of the craniofacial articulations. *Compendium of the American Equilibration Society*, **15**, 420

Moffett, B. (1984) *Diagnosis of Internal Derangement of Temporomandibular Joint, Vol. 1*, Continuing Dental Education Department, School of Dentistry, University of Washington.

Møller, E. (1966) The chewing apparatus. An electromyographic study of the action of the muscles of mastication and its correlation to facial morphology. *Acta Physiologica Scand.*, **69**, (Suppl. 280)

Morris, H. (1953) *Human Anatomy*, 11th edn, Blackistan Division of McGraw-Hill Book Company, New York

Munro, R. R. (1972) Coordination of the two bellies of the digastric muscle during basic movements of the jaw. *Journal of Dental Research*, **51**, 1663

Neff, P. A. and Suarez, M. D. (1983) Functional anatomy of the TMJ. *Ear, Nose and Throat Journal*, **63**, 9

Nevakari, K. (1956) Analysis of mandibular movement from rest to occlusal position. *Acta Odontologica Scand.*, **14**, (Suppl. 19)

Öberg, T., Carlsson, G. E. and Fajers, C. M. (1971) The temporomandibular joint. A morphologic study on human autopsy material. *Acta Odontologica Scand.*, **29**, 349

Ogus, H. D. and Toller, P. A. (1986) *Common Disorders of the Temporomandibular Joint*, Wright, Bristol

Penfield, W. and Rasmussen, T. (1950) *The Cerebral Cortex of Man*, The McMillan Co., New York

Piersol, G. A. (1923) *Human Anatomy*, 8th edn, J. Lippincot Co.

Pinto, O. F. (1962) A new structure related to the TMJ and middle ear. *Journal of Prosthetic Dentistry*, **12**, 95

Porter, M. R. (1970) The attachment of the lateral pterygoid to the meniscus. *Journal of Prosthetic Dentistry*, **24**, 555

Posselt, U. (1952) Studies in movements of the human mandible. *Acta Odontologica Scand.*, **10**, (Suppl. 10)

Prentiss, H. J. (1923) Regional anatomy, emphasizing mandibular movement with specific reference to full denture construction. *Journal of the American Dental Association*, **10**, 1085

Pröschel, P. (1987) An extensive classification of chewing patterns in the frontal plane. *Journal of Craniofacial Practice*, **5**, 56

Rees, L. A. (1954) The structure and function of the temporomandibular joint. *British Dental Journal*, **96**, 125

Rocabado, M. (1984) Diagnosis and treatment of abnormal craniocervical and craniomandibular mechanics. In *Abnormal Jaw Mechanics* (eds. Solberg and Clark) Quintessence Publishing Co., Chicago

Schumacher, G. H. (1961) *Funktionelle Morphologie der Kaumuskulatur*, Fischer Verlag, Jena

Sicher, H. (1951) Functional Anatomy of the temporomandibular joint. In *The Temporomandibular Joint*, (ed. B. C. Sarnat), C. C. Thomas & Co., Springfield, Illinois

Sicher, H. (1960) *Oral Anatomy*, 3rd edn, C. V. Mosby & Co., St Louis

Sicher, H. and Tandler, J. (1928) *Anatomie für zahnanzte*, J. Springer Verlag, Wien/Berlin

Silverman, S. I. (1961) *The physiology of the masticatory system*. C. V. Mosby & Co., St Louis

Steinhardt, G. (1958) Co-report: anatomy and physiology of temporomandibular joint: Effect of function. *International Dentists Journal*, **8**, 155

Symons, N. B. (1952) The development of the human temporomandibular joint. *Journal of Anatomy*, **86**, 326

Tryde G., Frydenberg, O. and Brill, N. (1962) An assessment of tactile sensibility in human teeth. *Acta Odontologica Scand.*, **20**, 233

Ulrich, J. (1896) The human temporomandibular joint. Kinematics and actions of the masticatory muscles. *Journal of Prosthetic Dentistry*, **9**, 399 (Tranlation by U. Posselt)

Vaughan, H. C. (1955) The external ptergyoid mechanism. *Journal of Prosthetic Dentistry*, **5**, 80

Vesalius, A. (1514–1564) A Flemish anatomist, scientist and physician. Illustrated dental occlusion

Vitti, M. and Basmajian, J. V. (1977) Integrated actions of masticatory muscles: simultaneous EMG from eight intramuscular electrodes. *Anatomical Records*, **187**, 173

Wallish, W. (1922) Das Kiefergelenk. *Zeitschrift für Anatomie und Entwicklungsgeschickte*, **64**, 533

Watt, D. M. (1981) *Gnathosonic Diagnosis and Occlusal Dynamics*, Praeger, Sussex

Widmalm, S. E., Lillie, J. M. and Ash, H. M. (1987) Anatomical and electromyographic studies of the lateral pterygoid muscle. *Journal of Oral Rehabilitation*, **14**, 429

Widmalm, S. E., Lillie, J. H. and Ash, M. M. (1988) Anatomical and electromyographical studies of the digastric muscle. *Journal of Oral Rehabilitation*, **15**, 3

Williams, P. L. and Warwick, R. (1980) *Gray's Anatomy*, 36th edn, Churchill Livingstone, London and Edinburgh

Williamson, E. H. (1980) Centric relation: A comparison of muscle determined position and operator guidance. *American Journal of Orthodontics*, **77**, 133

Williamson, E. H. and Brandt, S. (1983) Occlusion and TMJ dysfunction, *Journal of Clinical Orthodontics*, **15**, 333 (part 1); **15**, 393 (part 2)

Woelfel, J. B., Hickey, J. C. and Rinear, L. L. (1957) Electromyographic evidence supporting the mandibular hinge axis theory. *Journal of Prosthetic Dentistry*, **7**, 361

Zenker, W. (1954) Über die mediale portion des m. temporalis und dessen funktion. *Österr. Zschr. Stomat.*, **51**, 550

Index